THE AI CHATBOT FOR TEACHERS

300+ INSTANT CHAT PROMPTS FOR ENGAGING STUDENTS, SIMPLIFYING LESSONS, AND CUTTING DOWN ADMIN WORK

SHEILA SONNE

Copyright © 2024 by Halcyon Time Ltd

Published by: Halcyon Time Ltd

ISBN 978-1-80101-140-2

All rights reserved.

No part of this book may be reproduced in any form or by any electronic or mechanical means, including information storage and retrieval systems, without written permission from the author, except for the use of brief quotations in a book review.

Trademarks "ChatGPT" is a registered trademark of OpenAI OpCo, LLC. All other trademarks, service marks, and trade names mentioned in this book are the property of their respective owners. The publisher and author of this book are not affiliated with, nor have they been endorsed by, any of the products or vendors mentioned within this text.

The publisher and author have made every effort to ensure the accuracy and completeness of this book's content. However, they do not guarantee that the contents are error-free and expressly disclaim all warranties, including warranties of merchantability or fitness for any particular purpose. No warranty may be created by sales agents or in any written sales materials.

The mention of any organization, website, or product is for informational purposes only and does not constitute an endorsement by the publisher or author. The information provided by external sources is subject to change and may become outdated. The publisher and author are not providing professional services through this book. The strategies discussed may not be suitable for every individual or situation. Professional consultation should be sought where appropriate.

Neither the publisher nor the author shall be responsible for any potential damage or loss, including but not limited to loss of profits or other incidental, consequential, or special damages that may arise from the use of the information in this book.

CONTENTS

Introduction	vii
1. THE DAWN OF AI IN EDUCATION	1
AI 101: Unpacking the Alphabet Soup	3
A Quick History of Pre-ChatGPT Edu-Tech	4
Common Barriers to Adoption	6
I'm worried about privacy and data security with AI tools	10
Case Studies Demonstrating Easy ChatGPT Integration	10
2. GETTING STARTED WITH CHATGPT FOR TEACHERS	17
ChatGPT Basics: Class Is In Session	17
Setting Up ChatGPT in the Classroom	20
Customizing ChatGPT For Your Classroom	21
Helpful Add-ons and Plug-Ins for Educators	22
Prompting Hints	24
Basic Prompting Tips	25
Advanced Prompting Tips	26
Types of Feedback	27
What Can Go Wrong?	27
First Projects with ChatGPT: Simple Starter Activities	29
Icebreakers	29
Parent-Teacher Introductions and Communications	32
Warm-Up Activities	35
3. ENGAGING STUDENTS WITH CHATGPT	39
Assessing Individual Student Needs	40
Tailoring Lesson Plans and Materials	45
Providing Faster Feedback and Support	48
Encouraging Self-Guided Learning	55
Building a Community in the Classroom	60
4. BETTER CLASSROOM MANAGEMENT WITH CHATGPT	66
Making Teaching Admin a Breeze with ChatGPT	67
Making Behavior Management Less Stressful	73

5. AI IN SPECIAL EDUCATION AND INCLUSIVE PRACTICES	81
Meeting Diverse Needs	82
6. USING CHATGPT TO HELP DEVELOP IEPS (INDIVIDUALIZED EDUCATION PROGRAMS)	91
ChatGPT-Assisted IEP Development	92
Why It Matters	93
7. SIMPLIFYING MATH AND BUILDING STUDENT CONFIDENCE WITH CHATGPT	97
8. ENHANCING LITERACY AND READING ENJOYMENT WITH CHATGPT	105
How Much Is America Really Reading?	106
Screens Have Entered the Chat	106
Reading Makes a Difference	107
ChatGPT Tackles Literacy	107
9. MAKING SCIENCE COOL AGAIN WITH CHATGPT	117
10. BRINGING THE PAST TO LIFE WITH CHATGPT	124
11. DON'T FORGET ABOUT THE ELECTIVES	132
Art Class	132
Physical Education	134
Music Classes	135
Language Classes	136
12. ENCOURAGING CREATIVE PROJECTS AND CRITICAL THINKING	139
13. TRANSFORMING ASSESSMENT WITH CHATGPT	144
14. PROFESSIONAL DEVELOPMENT AND LIFELONG LEARNING WITH CHATGPT	150
15. MOST COMMON QUESTIONS TEACHERS HAVE ABOUT AI (AND YOUR ANSWERS)	158
What are your best tricks for catching AI-generated submissions?	158
What are some good AI detectors I can use to check my students' work?	159
How accurate are AI detectors and plagiarism detectors?	160
What's the difference between AI-generated and AI-assisted content?	160
What are the legalities regarding copyright and AI usage?	161
What are most students using AI for?	161

Are most schools developing AI policies?	162
What would an AI policy even look like?	162
Is it better to acknowledge AI in the classroom or let the students do their own research?	163
Will AI actually take over teaching?	163
Is ChatGPT suitable for all subjects and grade levels?	163
What other AI tools are out there besides ChatGPT?	164
Is it ethical to use AI in education?	164
16. THE FUTURE OF EDUCATION & NAVIGATING CHANGES WITH AI	165
17. SETTING UP YOUR OWN GPT	167
GPT Ideas	169
Important Considerations	170
Resources for Teachers	170
Afterword	177
About the Author	179
Works Cited	181

INTRODUCTION

Once upon a time, someone in a play said, "Those who can, do; those who can't, teach." Somewhere along the way, that phrase grew ugly horns and turned into "Those who can't do, teach." I'm not sure what kind of teachers these people must have been talking about, but it surely wasn't my colleagues. The education industry is brutal, and "those who can't teach" don't last very long. How could they?

At its core, teaching isn't about the money (there's barely enough to host one annual appreciation night for the staff in some districts, let alone keep the pencils stocked); it certainly isn't about indoctrinating children (there's barely enough time to each lunch most days, let alone run to the bathroom during 3-minute passing periods).

Teaching is a commitment to chaos. It requires untold patience. We are actors, continuously improvising to provide the stability our students need and deserve. We are pilots guiding our students through storms and unexpected turbulence despite pilot shortages left and right. We are mentors, magicians, and mechanics ready to lend our tricks of the trade and tools to the bright brains of our future.

Despite the economy, world conflicts, and noise, these children ARE the future, and it's our duty to get them there. We are tasked every day with the discernment necessary to know what's in their best interest and what's going to lead them astray. Educators are in a very different battle today than they were a century or even a decade ago. Many of us will remember the days before iPods and iPads, social media, and 24/7 internet access. That's just not how the world works anymore.

I'm not here to argue what's right or wrong, better or worse. It's just a fact—things are different. Teaching is no longer about the ABCs and 123s; neither is it confined to the four walls of a classroom. The outside world affects the classroom as much as the classroom affects the outside world. Those who "can't teach" will have a hard time adapting to this new world; those who love teaching might still struggle with the transition but will ultimately push on because that's what you do when you have dozens of tiny eyeballs looking up at you every day looking for an example of how to conquer this life.

The pandemic brought on a lot of talk about frontline workers, and if you ask me, teachers should be at the forefront of that list. If teachers were animals, we'd probably be an octopus with eyes in the back of our heads and eight tentacles getting pushed and pulled in eight different directions. We're all juggling the academic lessons with the life lessons with the state testing, grading, conferences, planning, and, for better or for worse, politics (no matter where on the spectrum you find yourself). The one thing we can all agree on is that we need a break.

What if I told you that a break was right around the corner or that you could catch one starting today? And what if I told you that this book has already laid the groundwork for you, so all you need to do is kick back your feet and decide what your new classroom will look like?

As teachers, we're in the business of honesty, so I'm going to level with you for a moment. This book isn't a magic wand that will fix all of the ails of the education system. It's not going to show you how to outsource 100% of your work to the machines or make your lesson plans so mundane that anyone could teach them. There are other books on these things out there, and you can find them if that's your prerogative.

Instead, here's what you can expect between these pages:

- The truth about individualized learning plans
- Secrets to manage your classroom without driving a wedge between your students or taking up too much instruction time
- How to make learning more accessible
- How to make math, reading, science, and history EASIER and MORE FUN
- How to guarantee job security despite rapid advancements in tech
- What you can do about AI plagiarism and cheating concerns
- What other teachers are asking about ChatGPT and the future of education

We'll journey through decades of education and see how ChatGPT is already tackling decision fatigue, challenging behaviors, and jam-packed schedules for teachers and students across the whole span of K-12 education in the US. And here's my promise to you: We won't focus on any of the blaring noise out there. We'll simply focus on the 100% actionable steps you can take as a teacher dedicated to ringing in a future that doesn't revolve around collective burnout, mass resignations, and parent-teacher conflict.

That's a big promise, I know, but I've seen it. Why ChatGPT? It's beginner-friendly and, best of all, free to use. It gives you more control over your classroom without the stress of having to stretch your budget even further or trying to find the time to learn something completely brand-new. If you know how to prompt it correctly (which you'll soon learn), ChatGPT is an unlimited source of classroom ideas and your new administrative assistant. It can take all the work you dread and make it 100x faster while making the work you enjoy 100x even more enjoyable.

Forget all the cookie-cutter advice and moral panic about AI bringing on the end of the world. It's all noise. Instead, break out your highlighters. You're going to want them. It's my hope that this book doesn't feel like work to you. It should feel like two colleagues chatting over coffee, reminiscing about how things were "back in the day," and realizing how good we have it now. Even if you only use FIVE of the 300+ prompts I've tirelessly created and tested, you should be able to save yourself a **minimum** of 3-5 hours per week. The math on that comes out to about 33 whole school days per YEAR.

One last request? Give yourself a pat on the back and celebrate your resilience. I don't care if it wakes your dog up from a nap. I want you to proudly fill your living room with these words: "I <u>am</u> THAT teacher. I <u>know</u> what's best for my students. I <u>deserve</u> a break."

Then, pour yourself a cup of something that makes you happy and journey on. We'll talk soon.

A Quick Note from the Author:

I tested each prompt in this book with ChatGPT myself to ensure they produced reliable and useful outputs. However, I did not use ChatGPT for any of the writing. It is up to us as educators to teach our students about the ethical limits of technology, and that is where I draw a hard line. Rest assured, this is a human-made book intended for other humans to read.

This book is also not intended to be political in nature whatsoever. If there are prompts you wouldn't consider using in your classroom, you are free to ignore them or tweak them as you see fit. I am not here to tell any educator how to run their classroom—I am simply here to serve as a bridge between an overworked past or present and a hopefully more manageable future.

You know yourself and your students best.

CHAPTER 1
THE DAWN OF AI IN EDUCATION

> "Learning is not attained by chance; it must be sought with ardor and attended to with diligence."
> —Abigail Adams, U.S. First Lady and Diplomat

THE CURRENT STATE of technology in the classroom gets a C-, maybe a C at best. Sure, we've got SMART boards and tablets in some classrooms, but others are still running on chalk, VHS tapes, and Elmo projectors. These disparities have always existed to some degree, so no surprise there, but comparing the technology of 2003 to what's possible now isn't even comparing apples to oranges; it's more like apples and orangutans.

Today's tech isn't about displaying bigger images or sharper text. It can think and talk! At least, it's as close as it can get to the real thing and only getting better by the day. Education now finds itself at a crossroads. Do we give this tech space in our classrooms or let it loom in the background like a very large, very pink elephant?

While some classrooms are diving into a wormhole leading straight to the future, others are sinking deeper into the past, trying to pretend this technology doesn't exist. They've opted for the "pink elephant" approach, and the children are paying the price. But the split might not be in the direction you'd think.

Forbes surveyed 500 teachers last year about their use of AI in the classroom. Do you have any guesses on how many teachers use AI for things like grading and classroom management?

1. 10%
2. 20%
3. 40%
4. 60%

The answer is D. Over half of the teachers surveyed had introduced artificial intelligence tools like ChatGPT into their classrooms. This group also unanimously predicted that AI will be used more widely in education over the next decade, not replacing but rather integrating with human instruction.

The Forbes study found that teachers aged 26 years and below were most likely to adopt AI in their classrooms. A concurrent survey from Inside Higher Ed found that even in classrooms where teachers did not use AI, over half of the students did.

Of course, good teachers are good critical thinkers, so the teachers voiced some concerns, too. These might be the same apprehensions floating around your mind right now: *What about cheating? What about ethics and integrity? Whose job will it be to introduce kids to ethical AI usage? Is this topic too controversial for a classroom?*

To ease your mind, let's take a step back and stroll down memory lane. As any good literature teacher would suggest, we must define the terms of our paper before we journey too far into the education jungle.

AI 101: UNPACKING THE ALPHABET SOUP

AI is defined as any "technology that enables computers and machines to simulate human intelligence and problem-solving capabilities" (IBM). This definition comes straight from the makers of the world's first supercomputers. You might also hear of "machine learning," which simply refers to the way that AI models are trained on the information.

We don't need to get into all the nitty gritty behind computer programming here, but now that you know the basic definitions, I want you to look at these two facts:

1. AI research has been around since the 1930s.
2. AI runs things like GPS, Siri, social media algorithms, and your Fitbit.

I think we can all agree that getting lost less and tracking fitness more have been generally good for humanity. Even social media has allowed international friendships and relationships to flourish despite some of the less great things that come with it. Ultimately, the definition of AI is so broad that it doesn't make sense to categorize it as "all good" or "all evil."

ChatGPT fits into the picture as an AI tool trained to understand human inputs and provide human outputs we can understand through conversational language. You don't need any coding experience to use it, and there are very few rules for how to interact with it. It is certainly not the first AI tool in education, nor will it be the last, but it IS currently the easiest to integrate. But again, we are getting ahead of ourselves.

Travel with me for a moment back in time.

A QUICK HISTORY OF PRE-CHATGPT EDU-TECH

Stanford University conducted a one-hundred-year study on artificial intelligence to observe patterns of AI use across various sectors. Their 52-page report notes that Intelligent Tutoring Systems (ITS) and Online Learning became prevalent far before the pandemic, which many mistakenly attribute to the rise in home-based learning. In fact, our own Air Force has used this technology to diagnose electrical systems problems in aircraft.

Heck, you've probably unknowingly used AI at some point during your teaching career. This is the same technology that instantly grades multiple-choice tests, and it's also the same technology that powers learning sites these days, like Duolingo for language lessons or Coursera and Edx for courses across dozens of disciplines. Or perhaps you've used quiz games like Kahoot to keep your students engaged?

Khan Academy is perhaps one of the most well-known resources in the teaching field, and even though the videos themselves aren't AI-made (as far as we know), you can bet the program itself is collecting data analytics through AI means. There's nothing frightening about that fact when you realize that this is the sole reason we've been able to increase education access tenfold over the last few decades. I know it might not always feel like it when you're a teacher down in the trenches or on the frontlines drowning in pressure related to assessments, but a lot of good has come from AI, and a lot of good will continue to come for those who learn how to integrate these technologies without relying on their districts for training and resources.

Amazon, Google, and Microsoft are pouring billions of dollars into AI research, and the U.S. government isn't far behind. In May of 2023, the Department of Education put out a 71-page document called "Artificial Intelligence and the Future of Teaching and Learning: Insights and Recommendations." If you'd like to read the full report, it will be linked with the resources at the end of this book.

Otherwise, here are the highlights:

- AI integration and a human-centric approach is best for educators and students
- AI tutoring can help the education system focus less on deficit-based approaches and more on asset-oriented approaches to learning
- AI can expand social and other aspects of learning in addition to more traditional types of learning like rote memory
- With the rise of behavioral issues across K-12, AI might have a place for teaching self-regulation and community-building
- While individualized lesson planning hasn't been realistic thus far, it's not so far away anymore
- AI can begin to limit bias seen across assessment and scholarship results
- AI must be trained on educator data, giving educators massive autonomy and power over the future

Tools like Quizlet, Socratic, and Otter.ai are great, but the problem is that many of these technologies still have barriers in the way of accessing them, whether it's a special invitation or a monetary fee. At least, that was true up until ChatGPT went public. ChatGPT is a tutor, teaching assistant, editor, and so much more wrapped up in one beginner-friendly platform, and *it's completely free.*

COMMON BARRIERS TO ADOPTION

If you're still unsure about this whole AI thing, I want you to know I hear you. But in my experience, 9/10 educators I talk to change their minds after being presented with the facts and being given the opportunity to interact with the technology themselves. That's the funny thing about teachers—we tend to have a knack for critical thinking.

I've talked to dozens and dozens of educators over the past year, with a combined total of a <u>millennium</u> of teaching experience, and this is what I kept hearing:

I'm just not good with all the tech stuff, and I'm too old to learn an entirely new system.

Although ChatGPT might be a "new to you" tool, you don't need to change anything about the way you talk or type with another human. That's the best part. ChatGPT can understand natural language, which means you can talk to it just as you would to a colleague, and it will respond in the same manner. Most people get the hang of it after just a couple of prompts. Besides, you don't even have to come up with the prompts yourself. That's what this book is for! It really is a "plug and chug" solution that you can get as creative (or not) as you want.

The resources in my district are stretched thin. I just don't know how my students will be able to access something like this.

Accessibility in education is always a valid concern, but both versions of ChatGPT give you instant access to more use cases than you could possibly imagine—even with a totally free account. ChatGPT can also run on phones, tablets, laptops, or desktop computers with an internet connection, making it readily accessible to many students. ChatGPT is also available in most libraries for students who don't have an internet connection at home. All in all, it's a nonissue. ChatGPT can help you create new lesson plans, discussion topics, etc., meaning that students don't even need direct access to the tool itself as long as you have it.

I'll also mention here that programs and grants are available to help schools improve their local community's digital infrastructure. Moreover, incorporating AI like ChatGPT can actually alleviate some of the resource pressures by offering personalized learning experiences without the need for extensive new materials or additional staffing.

AI is stealing people's jobs. I'm not supporting that crap.

The fear of job displacement by AI is understandable, and there will always be bad actors, but that's true of any technology or industry out there. The Internet brought on a hoard of cyber criminals and fraudsters, but are we going to ban the entire Internet because of that? Of course not. At one point, photography was feared to replace artists, but did that happen? Maybe for some, but the grand majority of artists simply **adapted** and **differentiated** themselves.

ChatGPT can perform administrative tasks, grade objective assignments, and provide teaching assistance, but that doesn't mean it can teach, mentor, or inspire students without a human counterpart. <u>It can't think for itself.</u>

It doesn't know Mason's favorite cereal or the fact that Tyler has been losing out on sleep because his family just welcomed a new baby. It doesn't know that Beth and Elizabeth have been best friends for ten years but recently got into their very first big fight because Elizabeth is moving away and Beth doesn't know how to process the news.

A teacher is so much more than some carefully crafted data or lesson plans and simply can't be replaced any way you look at it.
    AI might give the wrong information or biased content.

…As opposed to the news and the internet, which are always 100% reliable, right?

Like any educational tool, AI systems aren't infallible. It WILL mess things up, and it WILL get things wrong. If anything, this is the perfect moment in history to teach the bright minds of tomorrow critical thinking and digital literacy. We should encourage our students to question and verify information as much as we do ourselves.

There's one elementary school lesson of mine that will permanently be etched into my brain. It was the day we learned about the infamous "Tree Octopus." There was an entire website dedicated to the species with everything you could possibly want to know about the creature, including what it eats, where it lives, how long its lifespan is, and so on and so forth. There were even pictures! We were all so enamored that we couldn't help but tell all our families about this strange new creature at dinner that night. The only problem? It was a hoax.

The very next day, our teacher revealed that this was one of the greatest Internet hoaxes ever created (You can read about it more through the Library of Congress). It had been seen by thousands of other students and even grown adults who all fell for it without even batting an eye. We felt equally embarrassed that day, but of all the lessons my classmates and I learned that year, this is the one we continue talking about decades later because it stuck.

AI is inviting one of the greatest teachable moments of all time. It's this generation's Tree Octopus. At the same time, it can also be a tutor, study buddy, editor, and about one million and one other gadgets and gizmos. The cool part is if you ask ChatGPT to tell you about the Tree Octopus, it will clearly tell you it's a hoax as well as how it originated in the first place.

There's too much screen time already. We don't need more technology in the classroom.

We've all seen the effects of "iPad babies" and teens glued to their phones with our own eyes...and it's not pretty. Reduced attention spans, reduced distress tolerance, low emotional regulation. These have all been tied to too much screen time. The goal of integrating AI isn't to increase passive screen time any more than it has already increased over the past decade. Instead, tools like ChatGPT can be used to make screen time more interactive, meaningful, and personalized.

ChatGPT doesn't have any unlimited scrolling capabilities, pop-up ads, or endless streams of pictures and videos. It only responds to the prompts you tell it to with the exact information you requested. In a lot of ways, ChatGPT leads to LESS screentime because it gets you to the answer or solution you're looking for in record time.

In other words, what it's used for and what it outputs is really in your hands as an educator.

I'M WORRIED ABOUT PRIVACY AND DATA SECURITY WITH AI TOOLS

Teachers are scrutinized enough—believe me, I know this firsthand. The last thing we need is a data security scandal. Reputable AI tools comply with stringent data protection regulations and offer transparent privacy policies that you can read through directly on their site. Of course, most people don't have time to peruse fine print and legalese, but the most important thing to remember is that ChatGPT can't read any input unless you personally type it into its chat. This means it couldn't possibly know your students' full names, addresses, etc. unless you explicitly give it that information.

This list isn't exhaustive, but I hope it addresses some of your concerns regarding ChatGPT in the classroom. I'm aware you might still have them in the back of your mind as you continue through the book. All that I ask is that you keep an open mind and refrain from making any absolute judgments before reaching the back cover.

I also recognize that it might be difficult to understand exactly how this tool can be integrated into the classroom if you're not familiar with the technology, so I've put together three case studies to get your imagination going. These are an amalgamation of different K-12 educators I've met through the years and don't represent singular persons. Any school or name similarities are simply a coincidence.

Follow along with the case studies, or skip straight to Chapter 2 if you feel ready to jump into action!

CASE STUDIES DEMONSTRATING EASY CHATGPT INTEGRATION

Case Study 1: Elementary School Learning & ChatGPT

Background

Mrs. Thompson has been teaching for over thirty years and

currently teaches a 3rd grade class at Nelson Elementary School. She wants to find new ways to incorporate technology into her teaching but often feels behind the latest trends.

Challenge

Keeping her students' attention while also managing challenging classroom dynamics.

Barrier: Technological Unfamiliarity

Mrs. Thompson initially hesitated to integrate ChatGPT due to her lack of familiarity with AI technologies and her worries that the parents might not all be on board with the idea.

Solution

- **Grading Papers:** Mrs. Thompson used ChatGPT to help her with grading spelling and grammar tests. She thought the technology was surprisingly user-friendly and efficient, working exactly as planned from the start with minimal training. This process saved her a couple of hours of time every week, equating to a whole extra day of planning for the month.
- **Social Emotional Learning Plans:** She asked ChatGPT to help generate personalized age-appropriate SEL activities, most of which she hadn't considered before. She has noticed a considerable decrease in bullying after their first three lessons together.
- **Class Seating Charts:** With ChatGPT, Mrs. Thompson created a new seating arrangement that considered which students needed to be separated and which students would perform better when seated closer to her. She used initials to protect her students' data.

Overcoming the Barrier

- Mrs. Thompson watched some YouTube videos and bought a book just like the one you're reading to quickly catch her up to speed on the best practices for prompting.
- She started with simple ChatGPT applications, gradually exploring more of its capabilities as she became comfortable with the technology.
- Mrs. Thompson created a weekly newsletter to keep parents in on the loop about everything going on in the classroom, including upcoming exams, field trips, and AI education. Most parents appreciated the transparency and direct link between her quicker grading and their children's spelling success.

Outcome

- Mrs. Thompson noticed a sharp spike in the spelling test grades, which she attributes to the increase in turnaround time and instant feedback.
- The SEL lessons are engaging even the shiest children in the class, which has made for a more welcoming and collaborative classroom.
- The new seating chart has improved classroom dynamics and individual performance while alleviating the stress of planning everything herself in a classroom with a difficult layout.

Case Study 2: Middle School Learning & ChatGPT

Background

Mr. Lopez is a 7th-grade teacher at Beadle Middle School. He faces the challenges of keeping his students engaged in science while managing an increasing workload due to budget cuts and teacher shortages in his district. Historically, he knows that students tend to start struggling in his course around the midway point, which is arriving quickly.

Challenge
Balancing the administrative aspects of teaching with the need to prepare his students for the 8th grade.

Barrier: Resistance from Parents and Colleagues
Mr. Lopez encountered skepticism from parents and colleagues concerned about the reliance on AI for educational purposes, especially in math class, where there are clear right and wrong answers.

Solution

- **Problem Generation for Math:** Mr. Lopez used ChatGPT to generate new problems for his math examinations to counter cheating based on previous semester answer keys. He took the time to authenticate each new issue and solution.
- **Grading Papers:** He used ChatGPT for grading, freeing up time to host a study group for students who needed to retake the test.
- **Syllabus Creation:** After getting tired of answering the same questions about his retake and late work policies, Mr. Lopez used ChatGPT to generate a more robust syllabus with all his policies and FAQs.

Overcoming the Barrier

- Mr. Lopez organized an Open House to address concerns and showcase how ChatGPT was helping his students.

- He started with non-critical tasks, like brainstorming, to build trust and show tangible benefits, gradually expanding to more integral classroom activities, like grading.

Outcome

- After incorporating brand-new activities into the lesson, Mr. Lopez noticed an uptick in participation.
- Mr. Lopez's first study group was a resounding success, and he helped his students improve their grades. They even asked him to start regularly hosting study sessions, and he knows this will make a difference in getting his students ready for the next grade level.
- Mr. Lopez stopped getting so many questions about his policies because students had access to all the resources they needed in one place with their syllabus. Less classroom disruptions and less time answering the same question gives Mr. Lopez more time to focus on other, more important tasks.

Case Study 3: High School Learning & ChatGPT

Background

Dr. Lee teaches 11th-grade English at Horsetooth High School. She's passionate about preparing her students for college and "The Real World" through critical and creative thinking lessons and strong SEL support.

Challenge

Integrating advanced learning opportunities and comprehensive SEL into the higher-level curriculum in a way that keeps teenagers interested.

Barrier: Overwhelm with the Breadth of ChatGPT's Capabilities

Dr. Lee was excited about the potential of ChatGPT but felt overwhelmed by all the possible applications in her English class.

Solution

- **Grading Papers:** Dr. Lee used ChatGPT to provide initial feedback on essays, focusing on improving grammar and structure, while she provided deeper content analysis.
- **Social-Emotional Learning Plans:** She created customized SEL scenarios that were relevant to the literary themes they had been discussing in class and current events happening in the world around them.
- **Class Discussions:** ChatGPT simulated debates between all the students' favorite literary characters and posed challenging open-ended questions, challenging the students' worldview. This will prepare them to encounter differing opinions outside their school and community in the future.

Overcoming the Barrier

- Dr. Lee let the students lead the discussions and simply prompted ChatGPT with the things she knew they were interested in.
- Dr. Lee learned more about AI prompting from her students, who enjoyed talking to their favorite literary characters and practiced using the chatbot at home.

Outcome

- Dr. Lee's students stopped getting so many points taken off for grammar as their writing improved.
- The students grew more and more excited about college and appreciated the fact that Dr. Lee often incorporated bigger life lessons into her curriculum.

- Dr. Lee's class enjoyed their debates, even when they couldn't reach a consensus on some topics.

These case studies aren't all that fancy, but that's kind of the point here. **ChatGPT doesn't need to completely flip your existing lesson plans and teaching philosophy upside down.** Instead, you can incorporate it where it feels right.

The next chapter will show you how to get up and running in less than 5 minutes!

CHAPTER 2
GETTING STARTED WITH CHATGPT FOR TEACHERS

"An investment in knowledge pays the best interest."
—Benjamin Franklin

WHAT IF YOU could have a teaching assistant who knew everything and could teach anything at the drop of a hat? Without needing your supervisor's approval, without waiting for the administration to come around, and without spending hundreds of dollars out of your own pocket?

ChatGPT meet Reader, Reader meet ChatGPT.

Now that introductions are out of the way, we can get started with class.

CHATGPT BASICS: CLASS IS IN SESSION

OpenAI is the artificial intelligence community behind ChatGPT, the generative AI star of this book. They currently offer two plans to the public: a free plan and a premium plan priced at $20/month. What's the difference?

Premium users get access to a newer model and priority access during busier times, but the free version is perfectly adequate for most users. The other big difference is that the Premium version can access the Internet via plug-ins, while the free version of the model has a knowledge cutoff of January 2022 (an increase from its previous knowledge cutoff of September 2021). These knowledge cutoffs will continue to change as OpenAI puts out more updates.

Here's a handy visual that shows the differences better:

———

Feature	Free Version	Plus Version
GPT-3.5 Access and Usage Limits	Unlimited* Although the 3.5 model advertises unlimited use, some Free users have reported receiving error messages stating that they have reached their limit. If this happens, try logging out and logging back in.	Unlimited
GPT-4.0 Access and Usage Limits	N/A	As of January 5th, 2024, GPT-4 has a rate limit of 40 messages every 3 hours.
Model Capabilities	The ChatGPT 3.5 model can answer basic requests.	The ChatGPT 4.0 model can answer the most advanced requests and often shows more originality in its outputs.
Response Time	Standard response times, often slower during peak times	Priority response time, even during high-traffic
Access to Latest Features	Access to standard features only	Early access to new features and updates, like image or audio inputs and outputs.
# of Parameters the Model is Trained on	> 100 million	> 1 billion
Plug-Ins	N/A	Access to unlimited plugins and add-ons through the GPT store.
Advanced-Data Analysis	N/A	Standard
Browsing Capabilities	N/A	Standard
Image Generation	N/A	Standard

Author's Note: *You can use any of the plug-and-chug prompts included in this book with the free version, but you will not always be guaranteed the same results. There's never any commitment with the Plus version, so if you decide to upgrade and don't like it after a month, you'll only pay the $20. Do what works best for you and your class!*

SETTING UP CHATGPT IN THE CLASSROOM

If you've already got an account, you can go ahead and skip this section, but it's here for those who need it. ChatGPT tends to run better on desktops than mobile phones, though you can use it from any Internet-connected device. To use it on a desktop, you'll simply visit the OpenAI website to make an account. Once you have established an account, you can download the ChatGPT app from either the Google Play Store or Apple Store to enable mobile access.

Here's the step-by-step breakdown:

1. Type https://chat.openai.com/ into your browser.
2. Click "Create Account." It will default to the free account, but you can upgrade your account at a later time if you choose
3. Enter an email and password
4. Verify your email by clicking the link in the email that they sent you
5. That's it! You're officially ready to start prompting.

To interact with the model, you will simply type your prompt into the chat as a statement or question. You can then choose to respond and follow up on the initial output or create an entirely different prompt. You do NOT have to start a new chat each time you want to enter a new prompt.

Author's Note: *It's best to use a personal email rather than an educator email to ensure you retain access if you switch schools or decide to take advantage of ChatGPT's other features outside of education.*

CUSTOMIZING CHATGPT FOR YOUR CLASSROOM

I've already alluded to this, but ChatGPT has a "memory." If you are clear about your parameters from the start, you can save yourself the time and energy it would take to keep repeating yourself and reminding the chatbot of what exactly you're looking for. However, you will have to set this up in your account settings, or the "memory" will only work inside each individual conversation.

From a desktop:

1. Locate your profile in the bottom left corner of the screen and click on your name.
2. Click "Customize ChatGPT."
3. Provide your customized instructions for two questions:
4. What would you like ChatGPT to know about you to provide better responses?
5. How would you like ChatGPT to respond?
6. Click "Save" to lock in your settings.

The model will work best when you give it continuous feedback about what it's doing well and what you'd like it to improve. Essentially, it's another student!

Privacy, Safety, and Ethical Considerations

You should NEVER upload student identifiers to any online platform, including ChatGPT.

These are some common identifiers:

- Full Names
- Birthdays
- Addresses
- Emails
- Phone Numbers
- Student ID Numbers
- Photographs or Videos Displaying Their Faces

Other demographic information, like race, ethnicity, gender, and disability status, can also be a bit of a gray area if it points to one student in particular. Truth be told, I'm not sure why anyone would need to upload this type of information, but it bears noting, just in case. Basically, you should always utilize your common sense when you're crafting your prompts, and **when in doubt, leave it out**.

HELPFUL ADD-ONS AND PLUG-INS FOR EDUCATORS

OpenAI recently launched their "GPT Store," where creators can share their specifically trained models for free. There are already hundreds of add-ons and plug-ins to pick from specifically for educators. If you have a Premium account, you'll simply log in to your ChatGPT account as usual and then find where it says "Explore GPTs" in the top left corner from a desktop view. This will take you to the GPT Store, where you can give all the accessory models a try and save them for future use. You will also see the GPT's rating displayed based on others' interactions with it.

This is a small sample of GPTs that you might find helpful as an educator.

- ACT Prep: A GPT model that can help high school students prepare for the ACT.

- ADHD Companion: A GPT assistant that helps identify better-coping strategies.
- APA 7 Citation Helper: A GPT model that can help students with parenthetical and in-text citation formatting.
- AP Chemistry: A GPT model that can help students prepare for the AP Chemistry exam.
- Art Education Mentor: A GPT model intended to answer questions about various art forms and famous artists.
- Canva: A GPT model that can create posters, social media posts, and inspirational graphics.
- Chemistry Tutor: A GPT model that can help students with studying and data analysis.
- Elementary School Assistant: A GPT model created to help elementary school students understand their assignments and lessons better.
- English Tutor: A GPT model specifically created to help ESL students.
- IEP Assistant: An IEP writing and guidance tool.
- Grading Assistant for Teachers: A GPT tool that provides feedback for PDFs.
- Grammar Checker: A GPT model that fixes grammar mistakes across multiple languages.
- History: A specialized teacher model that can answer history-related questions.
- Math Solver: A GPT model that breaks down math problems step-by-step.
- MLA Citation Helper: A GPT model that can help students with parenthetical and in-text citation formatting.
- Multilingual Lesson Plan Creator: A GPT model that creates lesson plans for all ages and most subjects.
- Question Maker: A GPT that creates academic questions from PDFs.
- Rubric-Driven Grading Assistant: A GPT model that takes your rubric and compares it against student submissions.
- Rubric-Generator: A GPT model that can generate standards-based rubrics in seconds.

- Scholar GPT: A GPT model that analyzes papers from various scholarly sources.
- SEL Helper: A GPT model that recommends evidence-based SEL activities and lesson plans.
- Socratic Learning: A GPT model that can help lead discussions using the Socratic method.
- Spanish Tutor: A GPT model designed to help students practice their Spanish vocabulary.

Note: Because the GPT store is constantly changing, please vet each model on your own time and search for new ones, too! Skip to Chapter 17 to learn how to set up your own GPT!

PROMPTING HINTS

A "prompt" is anything you type into ChatGPT to produce an output or response. Now that you have this book in your pocket, you won't actually need to go out of your way to create your own prompts. They are all 100% tested and ready to plug and chug into the GPT model of your choice, but what kind of educator would I be if I didn't equip you with the skills and tools to forge your own learning path forward? **If you want to skip out on the prompting lesson, there will be no hard feelings. Skip to the next section!** Otherwise, I've put together a quick masterclass for you on basic and advanced prompting here.

One of the most common mistakes with first downloading ChatGPT is relying on a question format: "Can you…" "When was…" "How do I…" "Who is…" There's nothing inherently wrong with questions, and they can produce valuable outputs, sure, but you're missing out on ChatGPT's full capabilities if you only interact with the models this way.

ChatGPT has evolved to respond to statements and even fragments. Instead of questions, think more along the lines of: "Tell me a story about…" "Surprise me with a fact about…" "That doesn't seem right…" "I need something more like…" The models can even take on different characters, personalities, and voices for the entirety of a conversation, so you don't have to keep repeating yourself.

For the record, pleasantries and niceties like "please" and "thank you" certainly aren't required when interacting with the models, but if it feels more natural or you let one slip, it won't affect the output either way.

Types of Prompts:

- Specific Information Requests
- Idea Generation and Brainstorming
- Text Modification
- Creative Generation
- Text Extraction
- Q&A or Trivia
- Summarization
- Translation
- Parts of Speech Labeling
- Analysis

BASIC PROMPTING TIPS

Give the model specific parameters to work within:

- **Task** (what action should the model focus on?)
- **Context** (does the model need any background information?)
- **Word length** (# of words or paragraphs)
- **Tone** (casual, professional, playful, etc., or a mix of multiple!)

- **Persona or Style** (in the style of a particular genre, celebrity, etc.)
- **Copy format** (email, social media post, newsletter, IEP)
- **Words to include, words to avoid** (please include XYZ, please exclude XYZ)
- **Audience** (who will be interacting with the output)
- **Other constraints** (don't reference XYZ)

With a Premium account, the ChatGPT 4.0 model is capable of "memorizing" more context than the 3.5 model, allowing it to "learn" about your output preferences much quicker. This means you can often shorten your prompts as you become more familiar with the model and do not have to keep giving it the same background information within the same conversation.

You can also go into your settings and choose whether to toggle on the data controls. There should be a button for "Chat history & training." It will ask if you'd like to save your chats to your history and allow them to be used to improve the models. Any unsaved chats will be removed from the system within a month. You will need to toggle this setting on or off for every device you use with your ChatGPT account.

If you toggle this setting ON, an Activity Log will automatically populate and keep track of all of your conversations. It is often helpful to group conversations based on themes in order to easily join the conversation in the future without having to relay all of the same parameters. For example: You might have one chat dedicated to survey requests, another chat dedicated to science, another dedicated to parent communication, and so on and so forth. You have complete creative control!

ADVANCED PROMPTING TIPS

- **Multiple Perspectives:** Ask the model to provide answers from multiple experts in a single prompt

- **Expert Identity:** Give the model certain qualifications that might be helpful, i.e., "award-winning physicist" or "EdD honors student"
- **Use Scales and Proportions:** Instead of asking for a "more friendly" tone, try asking it for a 25% friendlier response

If the output isn't quite right, you can give the model feedback to try to finetune its subsequent outputs.

TYPES OF FEEDBACK

- **Corrections:** "You forgot to mention XYZ." "The formatting is off. I asked for XYZ instead."
- **Checks:** "Are you sure?" "Can you double-check that?" "That doesn't seem right."
- **Edits:** "Can you change the wording of the first paragraph?" "Make it longer." "Make it less cheesy." "Make it more relatable for our audience of XYZ."
- **Examples:** "Here's an example of what we need." "Please make it more like this."
- **Explanations:** "Can you explain XYZ to me like I'm in fifth grade?" "How do I do the second bullet point on the list?" "Can you expand on all the points in the outline?"
- **Additional Requests:** "Please write a blog outline based on the email." "Please turn this story into a poem." "Please add a paragraph about XYZ."

WHAT CAN GO WRONG?

ChatGPT might not always produce the outputs you expect because although it's trained on human data, there are still things about the human language that machines can't fully grasp. Don't be surprised if you come across the following every now and then:

- **Fluff and Filler:** Sometimes, chatbots say a lot without actually saying anything of value. The solution? Ask it to "cut out all the obvious."
- **Hallucinations:** If you ask ChatGPT for a list of mammals and a turtle appears on the list, you know the model hallucinated this fact. The solution? Ask it to "make sure there aren't any errors." Nine out of ten times, the model apologizes for the oversight and corrects itself.
- **Repetition:** The chatbot likes to get stuck on words like "leverage" and "crucial." As you get more familiar with the model, you'll pick up on these words and can ask the model to forgo them in future outputs.
- **Fake Statistics:** At times, the chatbot might provide statistics and sources that don't actually exist. Let it know that a mistake has been made whenever possible and always cross-reference other sources. Sometimes, this is simply a matter of its knowledge cutoff date. If a statistic changes after its cutoff, it will not be able to access the updated information on its own.
- **Conversation Interruptions:** When the servers get busy, the Chatbot might occasionally produce an error message as it tries to generate a reply. You can try to resubmit the request, but other times, you'll have to start an entirely new conversation. This is pretty rare, but it does happen.
- **Chatbot Refuses to Answer:** There are times that ChatGPT will tell you, "I'm unable to fulfill your request." More often than not, if you prompt it with "Why" or "Are you sure?" it will apologize and reset itself.
- **Chatbot Says It Cannot Interpret Your Data:** You may need to submit the information in a different format, i.e., as pasted text rather than an image.

Ready to try it? The next section includes several starter prompts to help you familiarize yourself with the ChatGPT model.

FIRST PROJECTS WITH CHATGPT: SIMPLE STARTER ACTIVITIES

ICEBREAKERS

Whenever it's time for new introductions, half of your classroom is likely feeling riddled with anxiety, and the other half is rolling their eyes. What if ChatGPT could make icebreakers a thousand times less painful?

ChatGPT Prompt Idea #1   Uncommon Icebreaker Activities

You are a seasoned [INSERT GRADE LEVEL] teacher. Brainstorm five fun icebreaker activities that will get the students excited for their new class. Do not list common icebreakers like "Two Truths and a Lie."

ChatGPT Prompt Idea #2   Introducing Yourself

What are some creative ways I could introduce myself to my new [INSERT GRADE LEVEL AND SUBJECT] class?

ChatGPT Prompt Idea #3   Active Icebreakers

What are some unique icebreaker activities that would get a group of [INSERT GRADE LEVEL] students out of their seats and moving?

ChatGPT Prompt Idea #4   Icebreakers for Shy Students

What are some icebreaker activities for a shy group of [INSERT GRADE LEVEL] students?

ChatGPT Prompt Idea #5   Icebreakers in Pairs

What are some fun icebreaker activities that my [INSERT GRADE LEVEL] students could complete in pairs in less than [INSERT #] minutes together?

ChatGPT Prompt Idea #6   Written Icebreaker Activities

What are some written icebreaker activities for a group of [INSERT GRADE LEVEL] students that don't require presenting in front of the class?

ChatGPT Prompt Idea #7   Recipe Card Activity

I have an idea for an icebreaker activity but don't have time to plan it. I think it would be fun for students to create recipe cards about their culture, background, hobbies, etc. Create a list of instructions I can share with them.

ChatGPT Prompt Idea #8   20 Questions

My class of [INSERT GRADE LEVEL] students would like to play 20 questions with you. Choose a random object and truthfully answer the next 20 questions. We will try to guess the object together. At the end of the 20 questions, let us know if we were correct or incorrect and what the object was. Please keep track of the questions.

Author's Note: You can also reverse roles and come up with an object together as a class, then have ChatGPT try to guess the object in 20 questions or less. The model might glitch due to the length of the conversation, especially if you are on the free plan. If you find that this consistently happens, try decreasing the number of questions or asking it to "try again."

ChatGPT Prompt Idea #9   What Do We All Have in Common?

I will upload three facts for every student in my class. I want you to analyze these facts and come up with five things that our class seems to have in common with one another. Please also generate some discussion topic questions based on those similarities.

ChatGPT Prompt Idea #10   Set a Timer

Please set a timer and message me in 30 seconds, letting me know that the time is up.

Author's Note: It would probably be easier to set a timer on your phone, so this prompt might not be the most useful, but it shows you just how many odd tasks ChatGPT is already capable of doing that most people wouldn't expect from a language model.

PARENT-TEACHER INTRODUCTIONS AND COMMUNICATIONS

Some of us aren't destined to be writers, and that's okay! ChatGPT can write wonderfully personal and professional letters, emails, and texts to students and their families.

ChatGPT Prompt Idea #11 . Parent-Teacher Introduction Letter & Email

You are a brand new [INSERT GRADE] [INSERT COURSE] teacher. You just graduated from the local university with your master's in education [OR INSERT OTHER FUN FACT ABOUT YOURSELF]. You are super excited about this new journey and want to write a letter to your new students and their families welcoming them for the school year. The letter should be less than 500 words and keep a cheery, casual tone while maintaining professionalism.

Please also include an email version.

ChatGPT Prompt Idea #12 . Edit My Email

I wrote my students' parents an email introducing myself, but I'm not sure if I like it. Can you take a look and suggest some edits to make it friendlier?

[PASTE EMAIL HERE]

ChatGPT Prompt Idea #13 . Parent Email Template

I need to create an email template for all my future communications with my [INSERT GRADE LEVEL] students' parents. It should include all my information: [INSERT NAME, PHONE NUMBER, COURSE NUMBER, ROOM NUMBER]

ChatGPT Prompt Idea #14 — Parent Survey

I would like to create an email survey for the parents of my [INSERT GRADE LEVEL AND SUBJECT] students to gauge their biggest concerns for the upcoming quarter. The survey should be no more than five questions long.

ChatGPT Prompt Idea #15 — SMS Template

I need to create some SMS templates for my students and their families. Create a short message for each of the following:

-Upcoming quiz
  -Upcoming test
  -The homework assignments for the week
  -New grades in the grade book
  -Upcoming field trip
  -Upcoming Parent-Teacher Conferences

ChatGPT Prompt Idea #16 — Welcome Sequence

I already sent the parents of my [INSERT GRADE LEVEL] students an introduction email, but I think I'd like to extend it into a welcome sequence instead. What ideas do you have for a month of emails that let the parents get to know me better?

ChatGPT Prompt Idea #17 , PTA/PTO Sign Up

I need to send my [INSERT GRADE LEVEL] students' parents an email letting them know that PTA/PTO sign-ups will be taking [INSERT DATE]. Please include these other pertinent details: [PASTE DETAILS]. The email should be less than 500 words and friendly but professional in tone.

ChatGPT Prompt Idea #18 , Chaperone Sign Up

I need to send my [INSERT GRADE LEVEL] students' parents an email letting them know that we are looking for volunteer chaperones for the following field trips:

[PASTE A LIST OF FIELD TRIPS]

Here's what they need to do if they'd like to sign up:

[PASTE RULES]

Parents can only sign up for a maximum of [INSERT #] field trips per year.

ChatGPT Prompt Idea #19 , Snack Duty

I need to send my [INSERT GRADE LEVEL] students' parents an email letting them know the rules of snack duty for our class. These are the rules: [I need to send my [INSERT GRADE LEVEL] students' parents an email letting them know]. Please draft an introductory email with these rules and a secondary email that will address the family chosen for snack week a week before their turn.

Please make parents aware of the following allergies in the classroom: [INSERT ALLERGIES]

ChatGPT Prompt Idea #20 . Illness Alert

I need to send my [INSERT GRADE LEVEL] students' parents an email letting them know that [INSERT ILLNESS] has been going around the school lately. Please also include any preventive measures they can take to safeguard their student and family. Also include our sick policy, which is as follows: [PASTE SICK POLICY]

WARM-UP ACTIVITIES

If you're anything like me, you've spent an embarrassing amount of time one month trying to craft the "perfect" warm-up activities, only for them to fall completely flat. (Who knew that a group of middle schoolers wouldn't appreciate a carefully crafted PowerPoint presentation filled with turtle puns?)

Anyway, that's enough about me and my poor attempts to make tweens laugh. There's some science I'd like to share with you. It turns out that students of all ages like having choices and autonomy, even if the choice they're making is seemingly insignificant, like which 5-minute warm-up activity to complete at the beginning of class. In fact, according to the National Library of Medicine, giving students choices has been proven to reduce challenging behaviors among school-aged children more than choosing for them. And that's something every classroom can do with less of.

Here's to hoping these choices can help:

ChatGPT Prompt Idea #21 . Unique Warm-up Activities

Create five choices for a five-minute warm-up activity for [INSERT GRADE] students enrolled in [INSERT CLASS]. The topic of the week is [INSERT LESSON THEME].

ChatGPT Prompt Idea #22 . Brainstorming Appropriate Debate Topics

Create a list of 10 debate subjects appropriate for a [INSERT GRADE LEVEL AND SUBJECT] class. The theme should be ethics.

ChatGPT Prompt Idea #23 . Reverse Pop Quiz

Help me create a reverse pop quiz warm-up activity, where my class of [INSERT GRADE LEVEL] students have to create their own pop quiz questions about [INSERT TOPIC] to demonstrate mastery. They will get to choose their own questions but will also need to provide the correct answers for each.

ChatGPT Prompt Idea #24 . Dice Warm-Up

How can we incorporate dice into our warm-up activities for a [INSERT GRADE LEVEL AND SUBJECT] class?

ChatGPT Prompt Idea #25 — Active Warm-Up

What are some warm-up activities for a [INSERT SUBJECT AND GRADE LEVEL] class that would get the students out of their desks and moving around the classroom?

ChatGPT Prompt Idea #26 — Bingo Warm-Up

Please create a BINGO board for a [INSERT GRADE LEVEL AND SUBJECT] class about [INSERT TOPIC]. It should have a Free Space in the middle.

Then, you can proceed with this prompt as a follow-up:

Please create [INSERT #] iterations and also include a caller card with all the potential words.

ChatGPT Prompt Idea #27 — Emoji Story

Please generate 5 emojis that could relate to yesterday's lesson about [INSERT TOPIC]. Instruct the students to create a story about these emojis to demonstrate their handling of the course material.

ChatGPT Prompt Idea #28 — Student Autonomy

What are some ways I can give my [INSERT GRADE LEVEL AND SUBJECT] class more autonomy during our warm-up exercises?

ChatGPT Prompt Idea #29   Students As Teachers

I need to design a warm-up activity. I think it would be fun for students to pretend to be the teacher and create their own 5-minute lesson plan about [INSERT TOPIC] to share with the class. Create the rules and instructions for me.

ChatGPT Prompt Idea #30   Cool-down Exercises

What's a fun cool-down activity my [INSERT GRADE LEVEL AND SUBJECT] students could do during the last 5 minutes of class each day to end class on a good note?

CHAPTER 3
ENGAGING STUDENTS WITH CHATGPT

> "The function of education is to teach one to think intensively and to think critically. Intelligence plus character– that is the goal of true education."
> —Martin Luther King Jr., Civil Rights Leader

TEACHING WAS ALREADY plenty difficult before iPods, iPads, and iPhones came along, but now, it's a constant battle for dwindling attention spans. The grade level doesn't matter; more and more students are bored, struggling, and would rather be elsewhere. And worse yet, some of these students don't hide that fact very well either —they make it known. Eyes rolling, constant yawning, restless legs, we've all seen it.

Giving our students the benefit of the doubt, this world is a roller coaster of constant global crises, bad news, and chaos. Of course, there's plenty of good and hope in the world, too, but it's funny how that chaos very rarely stays outside of the classroom for very long. But at the end of the day, that's the fight we chose.

Watching learning happen is like seeing the Disney fireworks for the first time. Pure magic. It's the one thing that keeps us all going. Officially, we might not have favorite students, but there's nothing like watching a D or a C slowly come up to a well-deserved B+ over the course of a few weeks and watching the quiet ones lead their first discussion.

Now, I'm not saying that ChatGPT is a wizard, but I've seen it produce magical results with even the most stubborn classroom of students. What's that phrase again? "If you can't beat 'em, join 'em." Students like tech, and that's what gives it superpowers. By meeting them where they're at, we show them we are invested in them as individuals and we care about them outside their academic success. A happy byproduct of that, most of the time, is that academic success comes anyway.

Here are some ways ChatGPT can turn the slackers into self-starters (or at the very least get to know what's behind the lack of participation):

ASSESSING INDIVIDUAL STUDENT NEEDS

The National Center on Teacher Quality compiled a list of surveys over the past few years that showed planning time is directly correlated with burnout, and the less planning time a teacher has, the more vulnerable they are. I'd guess that most of us have already been there at one point or another. Creating dozens of individualized lesson plans isn't all that realistic in any school, but ChatGPT makes it feel a little less impossible. Some educators might not agree with me here, and that's fine.

When we're dealing with limited resources, burnout, and assessments, we're bound to miss things. It turns out teachers are humans, after all. But jokes aside, I know what it's like to get caught up in the frustration of seeing students start falling by the wayside and feeling helpless to do anything about it or feeling wracked with guilt at not having noticed it earlier in the quarter.

As much as having a general lesson plan helps, we all know that certain groups of students and individual students will occasionally stray off the path. What if ChatGPT could serve as your crystal ball and predict these problems before anyone gets too lost?

Consider these three use cases:

- Conducting Initial Assessments
- Using Continuous Monitoring to Tweak Lesson Plans and Teaching Styles
- Giving and Receiving Honest Feedback

One great way to improve your teaching approach is to use ChatGPT to survey your students at the beginning of each semester to learn more about their current knowledge levels, learning styles, and interests. The introductory survey will serve as your baseline. As the quarter progresses, with ChatGPT's help, you can whip up other check-ins to gauge how your students are progressing and adjust your teaching approach accordingly.

You might have heard the Maya Angelou quote, "People will forget what you said. People will forget what you did. But people will never forget how you made them feel," and nowhere is it truer than a classroom. Open communication makes <u>all</u> the difference. If students feel comfortable sharing what's hindering their learning, it becomes easier to address those roadblocks. ChatGPT can create anonymous surveys to gather honest feedback from your students about the course material and pace or even the classroom dynamics.

Author's Note: Later chapters will include prompts for specific school subjects. Use the table of contents to jump right to the section you need! Chapter 13 also covers assessments more in-depth.

ChatGPT Prompt Idea #31 . Student Confidence Survey

You are an expert [INSERT GRADE LEVEL] [INSERT SUBJECT] teacher starting a new quarter. Create a survey that can help assess a new group of students' comfortability with [INSERT LESSON THEME OR SUBJECT].

ChatGPT Prompt Idea #32 . Learning Styles Survey

You are an expert [INSERT GRADE LEVEL] [INSERT SUBJECT] teacher who would like to get an idea of the students' preferred learning styles in your class. Create a survey based on the VARK model.

**Author's Note: VARK refers to Visual, Aural, Reading/Writing, and Kinesthetic.*

ChatGPT Prompt Idea #33 . Student Feedback Survey

You are an expert [INSERT GRADE LEVEL] [INSERT SUBJECT] teacher who would like to get student input on your teaching style.

Create a survey where students can be honest about the ways that your teaching may be helping or hindering their progress.

ChatGPT Prompt Idea #34 — Quarter Preview

I would like to assess my students' interest in the topics we are going to cover over the next [INSERT TIMEFRAME]. These are the topics: [INSERT TOPICS]. Create a quick survey to help me figure out what they are most interested in.

ChatGPT Prompt Idea #35 — Anonymous Feedback

We are watching a new video in my [INSERT SUBJECT] class today, but I'm not sure the students will like it. Help me come up with an anonymous survey where my students can share their honest thoughts about whether they'd like to watch these videos again.

ChatGPT Prompt Idea #36 — Mood Survey

I need to create a mood survey for my class of [INSERT GRADE LEVEL] students to give at the start of each week. It should ask where they stand on a scale of 1-10 for their overall mood. It can also include some optional questions at the end.

ChatGPT Prompt Idea #37 — If You Could Change One Thing

I'd like to give my students the opportunity to submit feedback. Help me create a survey based on the prompt, "If you could change one thing about our classroom, it would be..."

Include questions that revolve around several categories like social/emotional, resources, and academic achievement.

ChatGPT Prompt Idea #38   *What Are You Most Proud Of?*

I'd like to create a survey for my [INSERT GRADE LEVEL AND SUBJECT] class that allows them to brag about themselves and their progress so far this quarter. It should take less than 10 minutes to fill out.

ChatGPT Prompt Idea #39   *Course Progression Survey*

You are an expert [INSERT GRADE LEVEL] [INSERT SUBJECT] teacher who would like to measure how your classroom's anxiety related to [INSERT THEME] changes over the course of the quarter. Create a survey that can be given at the beginning, halfway point, and end of the course.

ChatGPT Prompt Idea #40   *Survey Analysis*

[Enter the survey data manually.]

Analyze this survey data, noting any relevant themes or patterns. Provide statistics when possible.

TAILORING LESSON PLANS AND MATERIALS

Now, consider these additional use cases:

- **Mountains of Metaphors:** ChatGPT loves creating metaphors, albeit cheesy ones. These can help simplify dense subjects that students might otherwise struggle to comprehend.
- **Learning Style Flexibility**: Learning styles are a controversial topic in education, but regardless of where you stand on the issue, ChatGPT can help you brainstorm new ways to bring your lesson plans to life. By analyzing your students' performance, ChatGPT can suggest new teaching methods and materials to help them better engage with the material.
- **Adaptive Learning Materials**: By inputting information about your students' progress, ChatGPT can quickly adapt your lesson plans to slow or speed up based on cold, hard data. This way, you can ensure that your students are always learning at the right pace for them.
- **Extra Credit Opportunities:** ChatGPT can instantly brainstorm dozens of ideas for extra credit projects that are more engaging than the typical essay (though it can come up with great writing subjects, too, as you'll see later on).

ChatGPT Prompt Idea #41 — Simplifying Metaphors

My [INSERT GRADE] students are struggling to understand [INSERT SUBJECT]. Create five metaphors about this topic that even a Kindergartener could understand.

ChatGPT Prompt Idea #42 — Customized Learning Style Plans

My [INSERT GRADE] students seem to do best with visual learning. How can I incorporate this into the lesson plans on the topic of [INSERT THEME]?

ChatGPT Prompt Idea #43 — Challenge Your Students

My [INSERT GRADE] students aren't being challenged enough. What are some activities they might enjoy for a lesson about [INSERT SUBJECT]?

ChatGPT Prompt Idea #44 — Unique Extra Credit Projects

Brainstorm a list of 5 extra credit projects of varying degrees of difficulty for [INSERT GRADE LEVEL] students on the subject of [INSERT TOPIC]. These projects should take no more than [INSERT TIMEFRAME] to complete and will need to be completed [INDIVIDUALLY, WITH A PARTNER, or WITH A GROUP].

ChatGPT Prompt Idea #45 — DIY Extra Credit

I would like to give my [INSERT GRADE LEVEL] students the opportunity to design their own extra credit projects. I need to create a submission form that outlines their project idea and explains how the project relates to these core competencies: [INSERT COMPETENCIES]

ChatGPT Prompt Idea #46 — Slowing Down

My [INSERT GRADE LEVEL AND SUBJECT] students seem to be confused by [INSERT CONCEPT]. Can you help me adapt my current lesson plan to slow down the pace without getting us too far off track?

[INSERT LESSON PLAN]

ChatGPT Prompt Idea #47 — Analyze Test Scores

I am about to share the test scores of my [INSERT GRADE LEVEL AND SUBJECT] students from the beginning of the quarter to now. Analyze this data and let me know what patterns you see. Do you think our course is moving at the right pace?

ChatGPT Prompt Idea #48 — Differentiating Lesson Plans

I have a lesson plan about [INSERT TOPIC] for my [INSERT GRADE LEVEL AND SUBJECT] class for next month, but I'd like to differentiate it based on no familiarity, some familiarity, and extreme familiarity with the topic so I can challenge my students appropriately.

Here's my current plan:

[PASTE CURRENT LESSON PLAN]

ChatGPT Prompt Idea #49 — Teach Me About the Learning Styles

I am a [INSERT GRADE LEVEL AND SUBJECT] teacher who would like to learn more about the learning styles and how they might manifest in my classroom. Act as a child psychologist and share evidence.

ChatGPT Prompt Idea #50 — Failing Rate

I am including the data for the number of students who have failed my course over [INSERT TIME PERIOD]. Help me understand some underlying reasons why this might be happening and any trends you notice. Ask me any clarifying questions you need to gather appropriate context.

The chatbot will then ask you several questions that you will need to answer truthfully to get to the root of the problem and can produce several individualized solutions.

PROVIDING FASTER FEEDBACK AND SUPPORT

Most of us have had to deal with a teacher or boss who doesn't seem to bother giving us actionable feedback, and how does that usually end? In my experience, initial confusion turns into frustration, which, if left unchecked for too long, turns into resentment. The resentment then begins to affect performance even more, and it all turns into a vicious, ugly cycle that doesn't really benefit anyone.

Is it easier to give minimal feedback? In the short-term, sure, but the long-term cost is almost always relationships and even self-esteem. And as educators, we owe it to our students to be honest with them. If the work they turned in exceeded our expectations, they deserve to know. If we thought they could have done better, they're equally deserving of the truth.

Just like in the workplace, faster feedback is often tied to better outcomes, including improved relationships and achievement (Indeed, 2022). So, how do we balance that truth with the harsh reality of our schedules?

ChatGPT isn't a time machine, but it comes pretty close. Get ready to enjoy:

- **Quicker Grading:** Grading is often the bane of our existence, especially when you have dozens of papers to crunch towards the end of the quarter. ChatGPT can be trained to provide detailed feedback based on your requirements and unique voice. It sounds crazy, but the three-part prompt for doing so is down below! ***For the record, you will still need to read the papers and assign the appropriate grades yourself, but AI can cut the whole feedback and grading process in half with just a few nudges.
- **Citizenship Reports:** While grades are important, character is equally important. With ChatGPT's help, you can generate personalized reports of encouragement to students based on qualities outside of their academic skills or achievements. These reports can help motivate your students and encourage them to develop character traits that will serve them throughout their lives.
- **Personalized Support:** With ChatGPT, you can set up an interface where students ask questions related to the curriculum and get instant, personalized help or explanations. While setting this up can be more time-consuming and requires more technical know-how, it can be a great way to provide extra support to your students and ensure that they fully understand the course material.

Building your own GPT model is covered further in Chapter 17.
- **Tutoring Outside of Classroom Hours:** Sometimes, students need a little extra handholding, but we simply don't have the time to stick around the same topic, or we're not available outside of office hours. You can share quick prompts for students to get the extra support they need outside of the classroom so they can better understand the "why" behind your feedback or the lesson at hand.
- **Unlimited Samples:** ChatGPT can instantly create sample requests for you to share with your students, but it can also generate assignment samples on its own. This is a win-win for everyone in the classroom. You save time on writing, and future students still have A-level work they can aspire to match or F-level work they know to avoid through their own efforts.

ChatGPT Prompt Idea #51 Grading Papers and Providing Paper Feedback

Teaching ChatGPT how to grade papers will consist of three steps.

The first part of the prompt will be the same for each student. From there, you will need to let ChatGPT know what an appropriate grade for the assignment is, and you will also need to provide your reasoning, which typically comes from the assignment rubric. This is especially useful if you often use the same rubric throughout the quarter. If that's the case, you can keep a Word Document with several variations for each grade level to cut down on copy/paste times in the future. The last part of the prompt requires addressing some of the student's mistakes or any areas of opportunity.

When you put it all together, it should look like this:

Please provide 200-300 words of feedback for this [INSERT GRADE LEVEL] student about their paper using a professional, academic, and friendly tone. Address the student with first-person language.

This student earned a [INSERT GRADE LEVEL] for their assignment based on the following criteria: [COPY/PASTE THE APPROPRIATE SECTION OF THE RUBRIC].

Tell the student to pay more attention to MLA formatting guidelines in the future and ensure they are using at least 3-5 unique citations throughout their paper [OR INSERT OTHER UNIQUE FEEDBACK].

ChatGPT Prompt Idea #52 — Encouraging Students

I would like to start sending my [INSERT GRADE LEVEL] students words of encouragement every week so they know that I appreciate them and am proud of them regardless of what the gradebook shows. Draft a 300-word note focused on [INSERT TRAIT].

ChatGPT Prompt Idea #53 — Clarifying Content

This is a prompt to share with your students for clarification outside of classroom hours.

I need help understanding the feedback on my latest assignment. I got the following question incorrect: [INSERT QUESTION]. My answer was: [INSERT ANSWER]. Please explain what I got wrong in simple terms. Please ask me any clarifying questions you may need to understand the circumstances better.

ChatGPT will ask the students any questions it needs regarding things like rubrics and assignment requirements before attempting to break down the solution step by step. The students can then follow up with as many questions as they need to gain a better understanding. This works for all subjects, including math, science, and reading.

ChatGPT Prompt Idea #54   Feedback Template

I need to create a feedback template for my [INSERT GRADE LEVEL AND SUBJECT] class. It should allow me to plug in a [LETTER OR NUMBER] grade as well as the rubric

ChatGPT Prompt Idea #55   A+ Samples

I need to create a sample paper for my [INSERT GRADE LEVEL AND SUBJECT] students so they can see what a perfect scored assignment looks like. The instructions for the assignment are as follows: [INSERT INSTRUCTIONS]. The rubric for the assignment is as follows: [INSERT RUBRIC]. Attempt to produce a perfect scored assignment, and I will give you feedback to improve it.

You will then have to go back and forth with the chatbot until it produces a sample to your liking. However, if you save this chat, it should be able to produce future samples more easily within the same chat.

ChatGPT Prompt Idea #56 . Sample Request

I would like to create a request form asking some students if I can use their essays as a sample for future students to see what a perfect scored assignment looks like. The form should explicitly ask for their consent and let them know that they are free to decline.

ChatGPT Prompt Idea #57 . After Hours Tutor

This is another prompt you may consider sharing with your students.

I am working on my homework for [INSERT GRADE LEVEL AND SUBJECT]. We are currently learning about [INSERT TOPIC], but I don't understand [INSERT SPECIFICS]. Pretend you are an expert academic tour and help me gain a better understanding by asking me questions and showing me detailed examples.

ChatGPT Prompt Idea #58 . Help Me Improve My Writing

This is another prompt you may consider sharing with your students.

I just got my essay back, and I scored a [INSERT GRADE]. I lost points because [INSERT SPECIFIC REFERENCES TO THE RUBRIC]. How can I improve my writing using my rubric as a guide?

ChatGPT Prompt Idea #59 . Sorting Into Groups & Peer Grading

A teacher of mine used to have us swap our quizzes with a random person in class and then grade each other's work. You can easily implement this by having ChatGPT randomly assign grading partners and then display the correct answers on your monitor. As another measure against cheating, you can have two versions of the same quiz with different answer keys.

Please create two versions of a quiz about [INSERT TOPIC] for my [INSERT GRADE LEVEL AND SUBJECT] class. Each quiz should be [INSERT #] questions long and [INSERT DIFFICULTY LEVEL].

Based on these initials, please randomly assign Quiz A to half the class and Quiz B to the other half of the class. The initials are [INSERT STUDENT INITIALS]

Now, please sort the students into pairs to grade each other's quizzes. Members of Quiz A should be grading the work of students who took Quiz B.

Please display the answer keys for each quiz.

ChatGPT Prompt Idea #60 — Student Feelings About Feedback

I would like to learn how my [INSERT GRADE LEVEL AND SUBJECT] students feel about the feedback they've been getting. Help me create a survey that addresses things like the turnaround time and quality of feedback.

ENCOURAGING SELF-GUIDED LEARNING

To give our students the best shot at the "Real World," we need to encourage lifelong learning and equip them with a sense of self-efficacy. There are evidence-based ways to do just that, including studying and journaling.

As teachers, we often forget that there was a time when we were students and had to learn how to study, even if it came more naturally to some of us than others. Contrary to popular cartoons, we couldn't just put a textbook under our pillows and learn through the process of "osmosis."

Studying takes self-discipline, curiosity, and endurance. It is far more than rote memorization, though memory has its time and place as well, and it's made up of more than one skill that, when applied consistently, will automatically become a habit.

To encourage effective study habits, we can use ChatGPT to do the following:

- **Teach Active Learning:** Simply reading a textbook and memorizing information may help students on exams, but this will only benefit them in the classroom. To equip our students with active learning strategies, we can ask ChatGPT to help us continuously brainstorm new activities. For instance, we can encourage students to summarize information in their own words, collaborate with their peers, and apply their knowledge to real-world scenarios. These active learning strategies will help our students understand the material better and retain the information for longer periods of time (we all know how rough that can be coming back from Winter break).

- **Promote Curiosity:** Our students should feel empowered to ask us questions and explore their interests. With ChatGPT's help, we can actively facilitate this by implementing project-based learning. This approach gives students the autonomy to pursue the things that captivate them the most while also encouraging critical thinking, problem-solving, and creativity.
- **Build Endurance:** By emphasizing the importance of hard work and dedication, we can help our students become more resilient and better equipped to handle life's challenges. We can also encourage them to set long-term goals and work towards them gradually rather than seeking immediate gratification (a rarity these days).
- **Foster a Growth Mindset:** Schools can be guilty of going overboard with boxes and labels, but we should show our students that they can develop into anyone they choose as long as they work for it. We can also remind them that life's teachable moments are nothing to be embarrassed about.

How do we do that?

- **Create Study Guides:** Creating study guides used to take hours of work, but ChatGPT can put together a thorough packet in a matter of minutes. The best part is that you'll be able to use these study guides for several years, as long as the topic doesn't change, i.e., Ancient History, World Religions, Pre-Algebra, etc.
- **Teach Better Study Habits:** Students can prepare for quizzes and tests or practice more effective habits by asking ChatGPT for suggestions tailored to their grade level and school subjects.
- **Encourage Reflective Journaling:** Journaling is proven to promote metacognition and self-awareness, which are both directly correlated with self-efficacy. You can generate a list

of prompts at the beginning of the course and then let students choose one prompt to answer weekly for participation points. Otherwise, students can also generate unlimited prompts to use as home as part of their personal de-stress routine.
- **Collaborate with Goal Setting:** Many districts these days require students to set SMART goals throughout their K-12 journey. ChatGPT simplifies the process and can offer ideas outside of the typical merit-based goals we tend to see every quarter.
- **Keep Our Students Motivated:** Every teacher knows motivation tends to ebb and flow around school breaks. ChatGPT can help you plan for the lows well in advance and maintain energy levels throughout the school year.

ChatGPT Prompt Idea #61 – Study Guide Creator

Create a thorough study guide for [INSERT GRADE LEVEL] students to help them prepare for their upcoming [INSERT CLASS] test. Make sure it includes the following topics: [INSERT LESSON THEMES]. It should also contain a variety of study tools, like multiple choice questions, true/false questions, and open-ended questions with explanations for each.

You will need to copy/paste the output into a Word document to share with your students.

ChatGPT Prompt Idea #62 – Unlimited Journal Prompts

Create a list of 20 journal prompts for a class of [INSERT GRADE LEVEL] students to reflect on their learning journey. The prompts should require deep thinking while sticking to school-appropriate topics.

ChatGPT Prompt Idea #63 — SMART Goal Generator

Create a list of 10 potential goals for a [INSERT GRADE LEVEL] student and explain how each follows the SMART goal framework. Use a mix of typical and atypical goals.

ChatGPT Prompt Idea #64 — Build Better Study Habits

I need to study for my upcoming [INSERT SUBJECT] test about [INSERT SPECIFIC THEMES]. What are 5 study habits for [INSERT GRADE LEVEL] that can help me be successful? Be specific about how these study habits will help me.

In the past, I have tried [INSERT STUDY TACTIC], and it has [WORKED/NOT WORKED] well for me.

ChatGPT Prompt Idea #65 — Motivating Students

How can I motivate my [INSERT GRADE LEVEL] students to complete their assignments on time?

ChatGPT Prompt Idea #66 — Celebrating Achievements

Provide ideas for recognizing and celebrating the achievements of our [INSERT GRADE LEVEL] students that they would appreciate.

ChatGPT Prompt Idea #67 — Encourage Active Listening

How can I encourage active listening among my [INSERT GRADE LEVEL AND SUBJECT] students?

ChatGPT Prompt Idea #68 — Goal Setting

I would like to help my [INSERT GRADE LEVEL AND SUBJECT] students set some goals for the quarter that don't follow the traditional SMART goal method. What are some other types of goals that might be beneficial?

ChatGPT Prompt Idea #69 — Self-Esteem Journal

What are some journal prompts that would be good for [INSERT GRADE LEVEL AND SUBJECT] students to build their self-esteem and self-efficacy?

ChatGPT Prompt Idea #70 — Active Learning

What are some ways I can model and teach active learning for a class of [INSERT GRADE LEVEL AND SUBJECT] students?

BUILDING A COMMUNITY IN THE CLASSROOM

I'm sure you can think back on your own education, and a teacher or two stands out for all the right—or all the wrong—reasons.

My memory takes me back to a Spelling Bee in the third grade. I mistakenly mixed up "their" and "there," as so many adults still do. My face immediately flushed with embarrassment and anger. I was angry at myself for having spent weeks preparing for this exact moment and losing. At that age, I was absolutely a sore loser because I didn't know any better. What could have been a really beautiful teaching moment immediately turned into public humiliation when my teacher made a point to tell the class how silly I was and how I really should have known better. The entire class was laughing *at me* instead of *with me*. Whether she meant to disrespect me, that's how I felt at the moment, and whenever I feel shame as an adult, I'm immediately transported back in time to that day, even if only for a moment.

In the grand scheme of things, there are worse things students deal with these days, but as an educator myself now, I'd handle that situation very differently. I would wait until after the Spelling Bee and take an extra hour to go over homonyms without singling out any particular student. I might even split the class into groups of two and turn it into a game. The point is that our students might mess up; they might not always get along, but conflict and mistakes are a simple fact of life. There are bound to be coworkers and bosses they don't agree with in the future or projects they fumble, and how we model respect in the classroom determines how those future conflicts are going to play out.

ChatGPT can foster community through:

- **Assigning Class Roles:** Giving students "jobs" in the classroom makes them feel part of something bigger than themselves, which can motivate them to stay engaged. You might think this is more for the younger students, but ChatGPT can brainstorm creative roles for students of all ages in the classroom. Middle school science teachers could "employ" a Lab Assistant, high school teachers could "employ" Project Managers and elective teachers might need help from Tech Support.

- **Creative Group Projects and Discussions:** Students can generate unlimited project ideas and discussion questions at the click of a button or plan the timeline and presentation of their projects.
- **Introducing Guest Teachers:** ChatGPT can take on the personality of your students' favorite characters. Why not have them guest-star for a lesson or two?
- **Supporting Peer-to-Peer Learning:** ChatGPT can suggest peer review guidelines for written assignments or create quizzes for students to test each other regarding the course content, encouraging constructive feedback and collaboration.
- **Providing Specialized Resources:** ChatGPT can generate a list of resources specific to your region.
- **Creating a Culture of Curiosity and Critical Thinking:** ChatGPT can create problem-solving scenarios based on real-world events that the students can actually relate to, which ultimately leads to more concrete learning.
- **Addressing Social and Emotional Learning Needs:** ChatGPT can provide scenarios and prompts encouraging students to reflect on their emotions, empathy, and interpersonal skills. It can also offer strategies for managing stress, resolving conflicts, and building resilience.
- **Restorative Justice, Conflict Resolution, and Anti-Bullying:** Bullying has been on the rise across most age groups over the past few years, and some districts don't have the resources for counseling. These situations are extremely disruptive to learning and it can feel impossible to move forward, but ChatGPT does a surprisingly good job

of coming up with conflict management plans involving restorative justice and other methods.

ChatGPT Prompt Idea #71 — Classroom Jobs

What are some unique age-appropriate "jobs" I can assign my [INSERT GRADE LEVEL AND SUBJECT] students to keep them engaged in the classroom?

ChatGPT Prompt Idea #72 — Group Project Planning

We are doing a group project for our [INSERT GRADE LEVEL AND SUBJECT] class. The requirements are as follows: [INSERT RUBRIC OR INSTRUCTIONS]. The plan we have so far is [INSERT PLAN]. Please create a realistic timeline to keep us on track.

ChatGPT Prompt Idea #73 — Peer Review Guidelines

Please create some peer review guidelines for our [INSERT GRADE LEVEL AND SUBJECT] class. Basic rules should include maintaining respect and making sure any feedback or criticism is constructive.

ChatGPT Prompt Idea #74 — Real-World Scenarios

My [INSERT GRADE LEVEL AND SUBJECT] class is learning about [INSERT SUBJECT]. Create three real-world scenarios where this topic might be applicable and give the students an opportunity to work on their problem-solving skills.

ChatGPT Prompt Idea #75 . Restorative Justice

We are dealing with [INSERT GRADE] bullying. How can we implement restorative justice tools to work through this situation?

You can also insert the specifics of the situation you're dealing with; just be sure to protect your students' personal and private information by either using fake names or initials.

ChatGPT Prompt Idea #76 . SEL Lesson Planning

Create a one-year-long SEL lesson plan for [INSERT GRADE LEVEL] students, making sure it is developmentally appropriate. Include specific lessons and activities for each theme and follow CASEL guidelines.

Author's Note: CASEL stands for The Collaborative for Academic, Social, and Emotional Learning.

ChatGPT Prompt Idea #77 . Cyberbullying

I need to create an impactful lesson about the effects of cyberbullying that will get through to [INSERT GRADE] students. It should have lessons, videos, and activities that are age-appropriate and demonstrate the seriousness of the matter.

ChatGPT Prompt Idea #78 . Celebrity Guest

Pretend you are [INSERT CELEBRITY]. Your mission is to teach my [INSERT GRADE LEVEL] students a quick lesson about [INSERT THEME] today.

ChatGPT Prompt Idea #79 — Finals Week Stress Relief

Finals are coming up for our [INSERT GRADE LEVEL] class. What are some fun stress relief activities we could plan for the next 3 weeks?

ChatGPT Prompt Idea #80 — Movie Day

Our [INSERT GRADE LEVEL] class will be watching a movie next week. What are some ideas for a movie focused on the central themes of [INSERT THEMES]? The movies should be rated [INSERT RATING, i.e., PG, PG-13].

For movies rated PG-13, include a waiver that will need to be signed by the parents before the students are able to watch the movie.

ChatGPT Prompt Idea #81 — Group Reflections

Provide five questions for [INSERT GRADE LEVEL] students to reflect on after completing a group project that required [INSERT REQUIREMENTS].

ChatGPT Prompt Idea #82 — Modeling "I" Language

What are some good examples of using "I" language to de-escalate conflict for a class of [INSERT GRADE LEVEL] students? How can we practice this language together?

ChatGPT Prompt Idea #83   Getting Everyone Involved

What are some other ways I can foster community inside my classroom and make all the students feel like they belong?

ChatGPT Prompt Idea #84   Mental Health Resources

What is a good list of national and local mental health resources for parents and [INSERT GRADE LEVEL] students in [INSERT STATE OR PROVINCE]?

ChatGPT Prompt Idea #85   Curiosity

What are some ways I can build my [INSERT GRADE LEVEL] students' curiosity in [INSERT SUBJECT]?

CHAPTER 4
BETTER CLASSROOM MANAGEMENT WITH CHATGPT

"The goal of education is the advancement of knowledge and the dissemination of truth."

—John F. Kennedy, former President of the United States

THIS WILL COME as no surprise to most educators, but according to a National Survey of Educators in District, Charter, and Private Schools conducted in 2022, only 1/4 of teachers reported spending more than 10 hours providing direct, whole-class instruction in a single week. If you think it's because we're all just lazy, you're probably reading the wrong book.

The same survey found that nearly 75% of teachers needed to spend between 1-3 hours addressing student disciplinary issues, another 1-3 hours attending required meetings, another 2-6 hours delivering assessments, and another 2-6 hours splitting their time between professional development and supervisory duties. Then there's planning time, time spent communicating with parents, administrative tasks, and about a dozen other tasks that don't have to do with teaching itself but make teaching impossible without it.

Let's also factor in that over half of teachers report dealing with between 1-3 instances of classroom disruptions across <u>each </u>of these categories per week: student discipline, student questions or concerns outside the classroom, announcements, administrator issues or meetings, and personal matters.

I'm no mathematician, but that's a LOT of hours spent on classroom management. Are we really wondering why so many people are leaving the field? Finding a teacher's assistant or body clone is out of the question for 99.99% of us, so I've included some friendly recommendations from teacher to teacher.

MAKING TEACHING ADMIN A BREEZE WITH CHATGPT

ChatGPT can save the day when it comes to things like:

Streamlining Classroom Documentation

- **Syllabus Creation and Expectation Setting:** Every course needs a syllabus, but you no longer need to spend hours creating one. ChatGPT can put together a solid one with minimal direction or proofread your drafts to ensure you're not leaving anything important out.
- **Rules Documentation:** ChatGPT can put together a list of classroom rules that are appropriate for every grade level, and you can also put a twist on it by asking the model to write the rules in the form of a poem or acronym to make it easier for students to remember.
- **Class Logs:** You might not remember what your students were working on eight weeks ago, but ChatGPT can access that information instantly, so you don't have to dig through piles of paperwork or dozens of digital files.

Managing Classroom Resources and Schedules

- **Resource Lists:** Get ready to generate essential resource lists for the academic year in mere minutes, customized to fit the size and needs of your course. ChatGPT can also suggest resources, activities, and assessment methods tailored to the lesson objectives and student preferences. This saves a TON of mental energy and might also save you some money.
- **Conference Planning:** Parent-teacher meetings are stressful enough without having to worry about all the extra work that comes with them. Why not hand the work over to ChatGPT, which can fully take over the planning and even draft the reminders?

Simplifying Communication with Parents

- **Providing Regular Updates:** Along the same lines, ChatGPT can generate regular updates for parents, summarizing their child's progress, upcoming assignments, and any relevant classroom news.
- **Templates That Retain the Human Touch:** You can partner with ChatGPT to create templates for emails and letters that still leave room for personalization.
- **Preparing for Emotionally Charged Conversations:** Telling a parent that their kid is performing poorly in your class or even failing is never fun. ChatGPT can transform into a coach in seconds, helping you prepare for these challenging conversations through role-play scenarios and scripts.

Streamlining Lesson Planning

- **Lesson Generation:** ChatGPT can help generate lesson plan outlines based on curriculum goals, teacher inputs, and student needs, saving time in the planning phase (Name ONE teacher who wouldn't benefit from more planning time).

- **Lesson Plan Editing:** Maybe you've got your planning 75% of the way done, but you just need a little extra push over the finish line. ChatGPT can work with whatever you've got.

And last but certainly not least:

Organizing Student Data and Insights

- **Identifying Trends in Student Performance:** Think quantitative AND qualitative
- **Analyzing the Effectiveness of Teaching Strategies:** How did your class respond to the new pop quiz format? What's not working so well?
- **Generating Recommendations for Interventions:** ChatGPT can take on the perspective of an administrator or evaluator to keep your classroom moving forward.
- **Predicting Future Learning Outcomes:** The more historical data you have, the better you can predict patterns, such as students tending to struggle around the 4-week mark or needing to retake a certain test.

ChatGPT can provide personalized intervention recommendations and support differentiated learning by identifying patterns, such as consistent improvement in a particular skill or recurring difficulties with a certain topic. For instance, if we notice several students struggling with a particular concept, we can adjust our teaching methods to address these difficulties and offer additional resources to help them improve. On the other hand, if we notice that several students are consistently improving in a particular area, we can offer challenges and more advanced materials to keep them engaged.

ChatGPT can also help correlate changes in teaching approaches with shifts in student outcomes. For example, if we notice that a particular teaching strategy significantly improves student outcomes, we can incorporate it more frequently into our lesson plans with just a few clicks. Alternatively, if we notice a particular teaching strategy is ineffective, we can adjust or eliminate it altogether.

But enough of the chit-chat. Here are some prompts you can start using today:

ChatGPT Prompt Idea #86 . Syllabus Creation

We need to create a syllabus for [INSERT GRADE LEVEL AND CLASS]. It should include all the basic elements of a syllabus with the following rules for late or missing work, assignment extensions, plagiarism, and detention: [INSERT ANY RULES YOU HAVE IN A LIST FORMAT]. Also, be sure to include a Frequently Asked Questions section with the following questions and answers: [INSERT ANY FAQs YOU GET OFTEN WITH THEIR ANSWERS IN A LIST FORMAT].

ChatGPT Prompt Idea #87 . Classroom Rules

We need to create a list of classroom rules to display on our bulletin board at the front of the class. These rules should be easy to remember and age-appropriate for [INSERT GRADE LEVEL].

For a fun twist, you can ask ChatGPT to make the rules rhyme. This usually helps younger students remember the rules of the classroom better, and you can recite them together at the start of each week.

ChatGPT Prompt Idea #88 — Resource List

I need to create a resource list for the upcoming quarter for my [INSERT CLASS] of [INSERT GRADE] students. I will have [INSERT NUMBER] students. What do you think I'll need? Ask any clarifying questions you may need to gather context.

ChatGPT Prompt Idea #89 — Parent-Teacher Conferences

Parent-teacher conferences are coming up for my [INSERT CLASS] of [INSERT GRADE] students. I will need to arrange meetings with the parents of [INSERT NUMBER] students. Create a plan I can use to prepare for these meetings as well as two reminder emails for a month before and a week before the meeting. Use a casual and friendly but professional tone. Sign the letters as [INSERT YOUR NAME].

ChatGPT Prompt Idea #90 — Classroom Newsletter

I would like to start sending the parents of my [INSERT CLASS AND GRADE LEVEL] students a weekly newsletter updating them about exciting class updates and important reminders. Help me plan three months of content.

ChatGPT Prompt Idea #91 — Challenging Conversations

I need to notify the parents of one of my students that they are currently failing in my [INSERT GRADE LEVEL AND CLASS]. Act as my education coach and help me prepare for this challenging conversation.

ChatGPT Prompt Idea #92 – Comparative Analysis for Past and Present Courses

[INSERT TEST SCORE DATA FOR TWO OR MORE DIFFERENT TIME PERIODS]. How do the test/quiz/assignment scores from [INSERT DATE] compare to [INSERT DATE]? What might account for these changes? What do you predict will likely happen in the future based on this data?

ChatGPT Prompt Idea #93 – Lesson Plan Editing

I am going to share my lesson plans from last year for my [INSERT GRADE LEVEL AND CLASS]. I will need you to help me optimize them. Respond with "yes" if you understand.

[WAIT FOR A RESPONSE]

[SHARE YOUR LESSON PLANS]

Author's Note: Asking for confirmation is useful when you're uploading long-form content and will reduce the risk of the chatbot getting confused by your request.

ChatGPT Prompt Idea #94 – Low Test Scores

I noticed that the exam scores for the latest test about [INSERT TOPIC] were much lower than I expected them to be. Help me figure out what might have been the root cause and how I can help my *students get back up to speed. Ask any clarifying questions you may need to gather the appropriate context.*

ChatGPT Prompt Idea #95 — Teaching Strategies

You are an expert school psychologist. Ask me questions about my teaching strategies to better understand how I operate my class. Then, produce suggestions for new teaching strategies that I should implement to help my students be more successful.

MAKING BEHAVIOR MANAGEMENT LESS STRESSFUL

If there's one thing I've learned about behavior management as a teacher, it's that you can do everything "right" and still face the wrath of tinier human beings who succumb to their very big emotions. We know it's not entirely their fault, especially when you throw hormones into the mix during the middle school and high school years, but that doesn't make it any easier.

According to the National Center for Education Statistics, nearly 1/3rd of public school teachers struggled with student misbehavior and tardiness during the 2020-2021 school year. This was across the board, from teachers brand new to the field to those with decades of teaching experience. And without getting political, the pandemic has had lasting effects on our teaching and students, backed by facts supported by the NCEP:

The 2021-2022 school year saw classroom disruptions more than double, with increased disrespect towards teachers and staff, electronic usage, and rowdiness. At the same time, over half of public schools reported needing more support across mental health, SEL training, hiring, and behavior management.

If you've been inside any classroom in the last few years, you don't need me to tell you any of these things. You've felt it. I simply share the statistics to reaffirm the fact that we are all in the same boat and that boat has been traveling some very rocky, uncharted waters.

Listen, ChatGPT isn't going to replace school counselors, magically hire a dozen other aids, or make the uninvolved parents at home suddenly care about their child who's slipping through the cracks. It just won't. But what it CAN do is help you save your mental, emotional, and physical energy so you're not burning both ends of the rope when you're at your wit's end.

Take these five ideas and see what a difference they can make:

Customizing Behavior Tracking with ChatGPT

- **Behavior Logs:** ChatGPT can create developmentally appropriate behavior logs that make tracking and analyzing patterns over time easy. You can even include the perspective of several experts, including behavioral psychologists and school psychologists.

Implementing AI-Driven Reward Systems

- **Rewards Program:** ChatGPT can suggest age-appropriate prizes for good behavior, like stickers, notebooks, and pizza parties. It can also keep track of each student's point balance if you'd like to create an experience where they earn points for good behavior and then get to go shopping with them.

Driving Conflict Resolution

- **Conflict Resolution Guides:** ChatGPT can guide teachers and students in resolving common classroom conflicts using evidence-based interventions.

Monitoring Class Mood and Engagement

- **Mood Tracking:** ChatGPT can create quick check-ins with students about their mood and engagement, which can tip teachers off as to what to expect in class that day.

Providing Personalized Behavioral Support

- **Support Plans:** Based on behavior logs and teacher input, ChatGPT can suggest personalized support plans for students who need additional help with behavior management or craft communications to parents and other educators explaining the situation.

ChatGPT Prompt Idea #96 — Detention Slips

I need to create detention slips for my [INSERT GRADE LEVEL] students. The slip should include a reason for the detention as well as a place for a parent's signature, along with all the other usual elements of a detention slip.

ChatGPT Prompt Idea #97 — Classroom Participation Interventions

I have been struggling with participation in my [INSERT GRADE LEVEL AND SUBJECT] classroom. Do you have any evidence-based strategies that my students might respond to?

ChatGPT Prompt Idea #98 — Behavioral Psychologist

[INSERT PROBLEM] has been a big issue in my [INSERT GRADE LEVEL] class lately. Here's what I've already tried to do to address the problem [INSERT SOLUTIONS].

Pretend you are an expert behavioral psychologist and create a better plan of action.

ChatGPT Prompt Idea #99 — Connecting with a Difficult Student

I have a [INSERT GRADE LEVEL] student who hasn't been participating in class. They often disrupt the lesson and disrespect their peers, and I don't know how to get through to them.

ChatGPT Prompt Idea #100 — Behavior Logs

I need to create an appropriate behavior log system for a class of [INSERT GRADE LEVEL] students. There should be a teacher-only log as well as a log that gets shared with the student and their parents.

ChatGPT Prompt Idea #101 — Reward System

I want to reinforce positive behavior in the classroom. What would an appropriate reward system look like for a group of [INSERT GRADE LEVEL] students? Use developmental and behavioral psychology to back up your plan. The budget for this plan is [INSERT NUMBER AND CURRENCY].

ChatGPT Prompt Idea #102 . Daily Mood Survey

I need to craft a daily mood survey for my [INSERT GRADE LEVEL] class that asks how they're feeling. I want the survey to ask about their experience with homework the prior night, whether they feel well-rested, and if they're hungry. It should also consider other factors that might affect a student's performance or well-being.

ChatGPT Prompt Idea #103 . Plagiarism Agreement

I'd like to draft a plagiarism agreement for my [INSERT GRADE LEVEL] that outlines what plagiarism is, including self-plagiarism and other examples of academic dishonesty that may or may not include AI. The agreement should include the consequences for breaking the code of ethics (detention, principal referral, suspension, and expulsion), as well as a place for a student and parent signature.

ChatGPT Prompt Idea #104 . Classroom Constitution

Our [INSERT GRADE LEVEL] students are going to create a constitution for our classroom. Create a plan for this activity that helps the students exercise their autonomy, critical thinking, and creativity.

ChatGPT Prompt Idea #105 . Parent-Teacher Behavioral Communication

A student in my [INSERT GRADE LEVEL AND CLASS] has been displaying [INSERT BEHAVIORS] for [INSERT TIMELINE]. Interventions I have tried include [INSERT ATTEMPTS]. I need to draft a letter explaining to the parents that this behavior is not contributing to their student's success and will not be tolerated. These are future consequences the student might be facing if nothing changes: [INSERT CONSEQUENCES].

The letter should be no more than a page long. It should use a strong but professional tone and also convey empathy for the student's situation. It should ask the parents how I can support them while explaining that I and the student need their help and involvement.

ChatGPT Prompt Idea #106 — Classroom Transitions

Our [INSERT GRADE LEVEL] students tend to struggle with transitions between different activities. Your task is to create an evidence-supported plan to make these transition periods less chaotic and stressful.

ChatGPT Prompt Idea #107 — Getting to the Bottom of Things

My [INSERT GRADE LEVEL] students have been struggling with [INSERT BEHAVIOR]. I need to put a plan together to figure out the root cause of the issue. My intuition tells me that these behaviors [INSERT YOUR THOUGHTS/FEELINGS], but I am not sure.

ChatGPT Prompt Idea #108 — Classroom Shop

My [INSERT GRADE LEVEL] class earns points for good behavior. Each week, they can use their points to purchase things from the classroom store. I need help keeping track of everyone's points along with the store's inventory. These are the current point balances:

[PASTE THE STUDENTS' INITIALS OR A RANDOMLY ASSIGNED ID# ALONG WITH THEIR CURRENT POINT VALUE]

This is currently what is on sale in the classroom store and how many points everything costs:

[INSERT INVENTORY AND PRICES]

Please alert me when we start running low on supplies for any particular item.

ChatGPT Prompt Idea #109 Conflict Guides

These are the conflicts that seem to keep coming up for my [INSERT GRADE LEVEL] students:

[INSERT CONFLICTS IN A BULLET POINT LIST]

I need help creating developmentally appropriate guides for each conflict that clearly define what strategies my students can use to prevent these conflicts or what they can do to de-escalate them.

Please include a guidance counselor slip that they can fill out if they'd like to request to speak to the counselor.

ChatGPT Prompt Idea #110 - Individual Support Plan

I need to craft a support plan for one of my [INSERT GRADE LEVEL] students who have been displaying the following behaviors:

[INSERT BEHAVIORS]

After talking to the student's parents and administration, these are the actions we identified as next steps:

[INSERT ACTIONS]

Please assess these actions and suggest any additional measures that you would take as a licensed school psychologist.

CHAPTER 5
AI IN SPECIAL EDUCATION AND INCLUSIVE PRACTICES

> "If a child can't learn the way we teach, maybe we should teach the way they learn"
> —Ignacia Estrada, Late Performer

EVEN IF YOU **aren't directly involved with special education, I encourage you to read this chapter.** Access to education affects us all, and we have a duty to advocate for every student—even those who aren't ours.

It's no secret that while every student deserves an education tailored to their unique needs, district shortages often make that reality more and more difficult. According to the Institute of Education Sciences, 40% of public schools either struggled to fill or were never able to fill special education teaching vacancies during the 2020-2021 school year, and the number of disabled students has steadily increased for the better part of a decade.

The teacher-to-student ratio in special education classrooms continues to be a problem. Disabled students are entitled to personalized instruction by law, but the law can't exactly make teachers appear out of thin air. Individualized attention is paramount for students with special needs, yet the reality is that classrooms are often overcrowded, and teachers cannot provide the personalized attention they wish they could to individual students with their own respective needs and abilities.

I know I'm preaching to the choir here, but traditional educational models, with their one-size-fits-all approach, are ill-equipped to meet this demand, often leaving students with special needs to navigate a system that doesn't fully accommodate their learning styles or provide the support necessary for their success.

Moreover, the integration of technology in special education, while offering potential solutions, also highlights gaps in access and training. Not all schools have the infrastructure to support the latest educational technologies, and not all educators are trained to utilize these tools effectively. This disparity in access to technology further widens the gap between what is possible in supporting students with special needs and what is currently being implemented in classrooms across the country.

At the end of the day, AI isn't a magic wand, but if it can make these students' lives and education even 10% better, don't we have an obligation to follow through? Let me be clear in saying that using ChatGPT as an aid is NOT a replacement for accommodations or specialized care, but it is, at the very least, an avenue worth exploring.

MEETING DIVERSE NEEDS

Have you considered using ChatGPT to modify your lesson content into accessible formats for students with different learning needs? With

ChatGPT's help, you can easily convert your lesson content into simplified text for readers with dyslexia, audio formats for visually impaired students, or interactive content for students on the autism spectrum. This way, you can ensure that every student in your class has access to the same educational content regardless of their learning needs.

ChatGPT can also recommend and help integrate assistive technology solutions that support learning, like speech-to-text software, text-to-speech tools, and educational apps designed for special education needs.

ChatGPT is particularly helpful across these four facets:

- Adapting Content for Accessibility
- Enhancing Communication for Non-Verbal Students
- Providing Assistive Technology Solutions
- Creating More Inclusive Classroom Environments

ChatGPT Prompt Idea #111   ADHD Accommodations

What are some reasonable ADHD accommodations for [INSERT GRADE LEVEL] students that I might not have thought of?

ChatGPT Prompt Idea #112   Autism Accommodations

What are some reasonable autism accommodations for [INSERT GRADE LEVEL] students that I might not have thought of?

ChatGPT Prompt Idea #113   Sensory Accommodations, Part One

What are some ways I can make my [INSERT GRADE LEVEL] classroom more sensory-friendly for students with sensory sensitivities?

ChatGPT Prompt Idea #114   Sensory Accommodations, Part Two

Provide a list of activities that engage students of all abilities in a [INSERT GRADE LEVEL] class, including those with sensory processing challenges.

ChatGPT Prompt Idea #115   Sensory Accommodations, Part Three

Suggest some helpful tools and techniques for managing noise levels in a [INSERT GRADE LEVEL] classroom.

ChatGPT Prompt Idea #116   Deaf or Hard of Hearing Accommodations

What are some reasonable deaf or hard of hearing accommodations for [INSERT GRADE LEVEL] students that I might not have thought of?

ChatGPT Prompt Idea #117   Blind or Visually Impaired Accommodations

What are some reasonable blind or visually impaired accommodations for [INSERT GRADE LEVEL] students that I might not have thought of?

ChatGPT Prompt Idea #118 — Discalculia Accommodations

What are some reasonable dyscalculia accommodations for [INSERT GRADE LEVEL] students that I might not have thought of?

ChatGPT Prompt Idea #119 — Dysgraphia Accommodations

What are some reasonable dysgraphia accommodations for [INSERT GRADE LEVEL] students that I might not have thought of?

ChatGPT Prompt Idea #120 — Testing Accommodations

What are some reasonable testing accommodations for [INSERT GRADE LEVEL] students that I might not have thought of?

ChatGPT Prompt Idea #121 — Dyslexia Accommodations

What are some reasonable dyslexia accommodations for [INSERT GRADE LEVEL] students that I might not have thought of?

ChatGPT Prompt Idea #122 — Simplifying Instructions

I am going to provide you with the instructions for an assignment. I need you to simplify these instructions as an accommodation request for [INSERT GRADE LEVEL] students with [INSERT NEEDS].

[INSERT ASSIGNMENT INSTRUCTIONS AS TEXT OR AN IMAGE]

ChatGPT Prompt Idea #123 — Inclusive Classroom, Part One

What are some ways that I can make my [INSERT GRADE LEVEL AND SUBJECT] class more inclusive of students from different backgrounds and abilities?

ChatGPT Prompt Idea #124 — Inclusive Classroom, Part Two

How can I teach my [INSERT GRADE LEVEL] class about inclusive practices in a way that is sensitive and they will respond well to?

ChatGPT Prompt Idea #125 — Heritage

How can we celebrate [INSERT HERITAGE] more in our [INSERT GRADE LEVEL AND SUBJECT] classroom?

ChatGPT Prompt Idea #126 — Parent-Teacher Communication Regarding Learning Difficulties

I need to draft a letter to the parents of a [INSERT GRADE LEVEL] student. I have noticed that the student seems to be struggling with

[INSERT SKILLS]. I would like to discuss ways that we can all support this student over the phone. The letter should voice my concern while also remaining optimistic. It should also be 500 words or less and include a compliment to the student about their [INSERT SKILL OR TRAIT].

ChatGPT Prompt Idea #127   Tech Instructions

Our [INSERT GRADE LEVEL AND CLASS] students will be completing an assignment about [INSERT SUBJECT]. They will need to use [INSERT TECHNOLOGY] for this project. I need to draft up some basic instructions about how to make an account and use this technology at home. It should include basic troubleshooting instructions in case the student's family is not familiar with the technology.

ChatGPT Prompt Idea #128   Quiet Corner

I need help designing a dedicated corner of my [INSERT GRADE LEVEL] classroom for calming and emotion regulation. The area is approximately [INSERT MEASUREMENTS] in size. The tools and objects we currently have accessible are [INSERT TOOLS AND OBJECTS]. What else does our quiet corner need?

ChatGPT Prompt Idea #129   Sensory Bins

I need to plan five different activities for our sensory bins. They must be developmentally appropriate for [INSERT GRADE LEVEL] and safe for students of all abilities.

ChatGPT Prompt Idea #130   Student Narrator

Please read the following text out loud at a moderately slow pace and help me understand it: [INSERT TEXT]

ChatGPT Prompt Idea #131   Sign Language Basics

Teach me some basic sign language that may be helpful in communicating with students in a [INSERT GRADE LEVEL] classroom.

ChatGPT Prompt Idea #132   Communication Skills

What are some ways that I can work on my verbal and non-verbal communication skills to make my [INSERT GRADE LEVEL] classroom more inclusive?

ChatGPT Prompt Idea #133   Learning Disability Signs

What are some signs of a potential learning disability that may manifest in [INSERT GRADE LEVEL] students, and what can I do to help?

ChatGPT Prompt Idea #134   Training Ideas

What are some training opportunities that could help me strengthen my commitment to running a more inclusive and accessible classroom?

ChatGPT Prompt Idea #135 . Advocacy

How can I become a better advocate for my students regarding accessibility both inside and outside the classroom?

ChatGPT Prompt Idea #136 . Advocacy Leaders

Who are some advocacy leaders I can look up to within the education system?

ChatGPT Prompt Idea #137 . Book Recommendations

What are some books I could read about making education more accessible?

ChatGPT Prompt Idea #138 . Visual Menu

If I wanted to create a visual menu to help a non-verbal student communicate their needs, what types of images should I include? Please ask me any follow-up questions you may need to get a more thorough understanding. Act as a licensed para or disability specialist.

ChatGPT Prompt Idea #139 . Speech-to-Text Software & Text-to-Speech Tools

How would I go about integrating speech-to-text software and text-to-speech tools inside my classroom? Be as specific as possible and break it down for me step by step.

ChatGPT Prompt Idea #140 . Inclusivity Guide

How would I go about creating an inclusivity guide for my classroom that I can share with the parents of my [INSERT GRADE LEVEL] students? I want to address the fact that we will have students of varying abilities without singling any child out. I also want to provide tools for parents and students to have conversations at home about what it means to be a good citizen and a good friend.

CHAPTER 6
USING CHATGPT TO HELP DEVELOP IEPS (INDIVIDUALIZED EDUCATION PROGRAMS)

"Everyone who remembers his own education remembers teachers, not methods and techniques. The teacher is the heart of the educational system."
—Sidney Hook, American Philosopher

VARIOUS MENTAL HEALTH diagnoses like ADHD have been on the rise across schools in America to the tune of millions of students. CDC data from 2016-2019 estimates that 6 million children between the ages of 3-17 have received an ADHD diagnosis, with boys twice as likely to receive the diagnosis than girls. Comorbidities among those kids diagnosed are also very common, with nearly half the children experiencing a behavior or conduct problem, 30% experiencing anxiety, and smaller percentages experiencing depression or Tourette syndrome.

Of course, ADHD isn't the only diagnosis that needs attention, but these students often experience undue challenges in classrooms that don't always support their needs. It's easier to advocate for students when they have an established IEP, but getting the IEP in the first place is a different story. Every district has its own hoops to jump through, and parents aren't always on board either, even when their student would clearly benefit from the extra support.

It's not our place to overstep when parents and administrators have made their call, but we never cease to be child advocates, either. And with or without an IEP, we can all make that happen—especially with the help of ChatGPT.

CHATGPT-ASSISTED IEP DEVELOPMENT

Once you get the ball rolling, ChatGPT can help maintain that momentum through the following:

Gathering Student Performance Data and Insights

- **Data Compilation:** ChatGPT can continue to help you compile and analyze student performance data, leaving a paper trail that can advocate for your student.

Creating Templates for IEP Documentation

- **Template Generation:** ChatGPT can generate customizable IEP documentation templates that streamline the creation process, making sure no major components are missing, and all the bases are covered.

Encouraging Collaboration Among Educators, Specialists, and Parents

- **Collaborative Platforms:** ChatGPT can make sure the entire team is on the same page when it comes to supporting a student.

Streamlining the IEP Review and Update Process

- **Review Assistance:** You can automate the review and update process for IEPs, using ChatGPT to highlight areas needing adjustments and suggest modifications based on the latest student data.

WHY IT MATTERS

A middle schooler who wants to understand the reason behind something might ask, "What's the Big Whoop?" Kids say the darndest things, don't they? (ChatGPT can keep a log of your favorite phrases and turn it into a yearbook at the end of the academic year, by the way).

IEPs matter for a multitude of reasons:

Explaining the Purpose and Components of an IEP to Stakeholders

- **Educational Content Creation:** Teachers can use ChatGPT to generate emails and letters that explain the purpose, components, and importance of IEPs to parents, students, and new educators.

Personalizing Education Plans

- **Advocating for Individual Needs:** ChatGPT can analyze student data and learning patterns to present evidence in favor of developing an IEP and advice tailored to specific accommodation needs.

Efficiency

- **Shorter Wait Times:** ChatGPT can speed up the timeline of getting an IEP approved from months to weeks by streamlining the documentation portion of the process, assuming all the stakeholders actively participate in the process.

ChatGPT Prompt Idea #141 ⸱ Goal Setting

Based on the following skills and abilities, give me some specific and measurable goals for a [INSERT GRADE LEVEL] student to work towards: [INSERT SKILLS AND ABILITIES]

ChatGPT Prompt Idea #142 ⸱ Documenting the Need for an IEP

I am a [INSERT GRADE LEVEL] teacher. I think one of my students could benefit from an IEP. These are some things I've noticed: [INSERT TRAITS AND DESCRIBE THE CIRCUMSTANCES]. How can I document their learning journey?

ChatGPT Prompt Idea #143 ⸱ IEP Conversations with Parents

I will be meeting with the parents of one of my [INSERT GRADE LEVEL] students to discuss the possibility of an IEP for the following accommodations: [INSERT ACCOMMODATIONS]. How can I prepare for this difficult conversation to let the parents know I will support them and their students through this journey?

ChatGPT Prompt Idea #144 ⸱ IEP Was Denied

I am a [INSERT GRADE LEVEL] teacher whose student was denied an IEP. I think they could benefit from the following accommodations: [INSERT ACCOMMODATIONS]. What are some other ways I can continue to support them?

ChatGPT Prompt Idea #145 — Transition Planning

What are some transition planning considerations an IEP should include for a [INSERT GRADE LEVEL] student with the following accommodations in place: [PASTE ACCOMMODATIONS]

ChatGPT Prompt Idea #146 — IEP Timeline

I'd like to generate a realistic timeline to help one of my [INSERT GRADE LEVEL] students get approved for an IEP. The timeline should include all the documentation and stakeholders that are typically involved in this process, as well as estimates and suggestions for each step.

ChatGPT Prompt Idea #147 — IEP Advocacy

My school district has not historically accepted many IEPs. How can I advocate for my students and for IEPs, given the circumstances? Provide evidence in favor of IEPs and their success.

ChatGPT Prompt Idea #148 — IEP Case Studies

I am a [INSERT GRADE LEVEL] teacher, and I find that many students' parents don't quite understand how an IEP works or how it might be beneficial. Help me illustrate the power of IEPs through 3-5 case studies.

ChatGPT Prompt Idea #149 — IEP Collaboration

What are some tips for communicating with other teachers, therapists, and psychologists as an advocate for my student who is on an IEP?

ChatGPT Prompt Idea #150 — Adjusting IEPs

I have a [INSERT GRADE LEVEL] student who has put on an IEP. Their original accommodations are as follows: [INSERT ACCOMMODATIONS]. I have noticed the following are really helpful to the student: [INSERT BENEFITS]. On the other hand, I have noticed the following is not helpful to the student: [INSERT DIFFICULTIES]. Offer suggestions to the IEP and explain your reasoning.

ChatGPT Prompt Idea #151 — General IEP Template

Craft an IEP template that I can fill out for my [INSERT GRADE LEVEL] students tailored to their individual needs.

CHAPTER 7
SIMPLIFYING MATH AND BUILDING STUDENT CONFIDENCE WITH CHATGPT

"To teach is to learn twice."
—Joseph Joubert, French Essayist

I'VE NEVER TAUGHT MATH, but my heart goes out to those of you who do. According to various educator blogs, it consistently ranks as one of the most difficult subjects to teach next to foreign languages. And let's face it, math IS a foreign language for many of our students.

Unfortunately, the pandemic didn't do our students any favors. Assessment scores tanked across the board, and K-12 students had already been struggling with geometry and statistics for many years prior (Sparks). If that's not bad enough, the United States consistently falls behind other countries.

In 2020, the USA ranked 9th in Reading and 31st in math literacy compared to teenagers across 79 countries, following a trend that's been observed for nearly 20 years now (Richards). The big brains at Stanford have argued that the order and focus of most US curricula is the issue. Many schools separate Algebra I from Algebra II, with

Geometry in between, and there is also a pretty wide gap in teaching courses like Data Science or Coding, which are the norm in other parts of the world.

I'll reiterate that this book isn't the place for debates surrounding the Common Core curriculum, but any way you slice the cake, it isn't looking good. Our students and fellow educators are struggling.

ChatGPT isn't going to change the curriculum overnight, but it might be just the fulcrum that math teachers need to give their students an edge in the upcoming years, following the suit of countries like Estonia, which prioritize edutech across all age groups.

Take these three examples:

Transforming Math Lessons

- **Interactive Problem Solving:** ChatGPT can instantly create real-world math problems and display their solutions.
- **Visualizing Concepts:** At this time, AI photo generation is still a little wonky, and you can't always rely on it. However, ChatGPT can help students visualize complex mathematical concepts by describing them in further detail or recommending relevant resources.
- **Adaptive Learning Paths:** ChatGPT can customize math challenges to fit each student's skill level. Quick hack: Try setting new parameters by asking it to create a problem that is 25% (or some other percentage) more difficult.

Enhancing Understanding and Engagement

- **Gamification of Math Learning:** ChatGPT is filled with game ideas to make math more fun for students of all ages.
- **Collaborative Projects:** ChatGPT can facilitate group projects that get students excited about learning a subject they might be nervous about.

Preparing for Math Assessments

- **Personalized Revision Guides:** ChatGPT can create study materials tailored to individual needs.
- **Tests and Quizzes:** ChatGPT can also generate tests and exams, including multiple-choice questions, True/False questions, fill-in-the-blank questions, and open-ended questions.
- **Feedback and Improvement Plans:** ChatGPT can provide automated feedback on assessments and homework assignments to guide student improvement.

ChatGPT Prompt Idea #152   Pi Day Activities

Help me brainstorm a list of 5 fun age-appropriate activities for [INSERT GRADE LEVEL] to enjoy Pi Day.

ChatGPT Prompt Idea #153   Real-Life Applications

What are real-life applications of [INSERT MATH CONCEPT] that would connect with [INSERT GRADE LEVEL] students?

ChatGPT Prompt Idea #154   Study Prep

Create a series of word problems and multiple choice questions related to [INSERT TOPIC] to help a [INSERT GRADE LEVEL] student prepare for their upcoming [INSERT TYPE OF MATH] exam. Include the answers and a step-by-step explanation of how you arrived at each solution.

ChatGPT Prompt Idea #155   Scaffolding

We need to solve the following problem: [INSERT PROBLEM]. Take a scaffolding approach to explain all the relevant terminology applicable to the problem and how to arrive at the solution.

ChatGPT Prompt Idea #156   Group Project

Create five ideas for an engaging group project made for [INSERT GRADE LEVEL] students. The projects should have a deadline of [INSERT TIMELINE].
    Once you choose your favorite idea, proceed with the following prompts:

Create a rubric for the project that assesses the students on the following criteria: [INSERT CRITERIA].

ChatGPT Prompt Idea #157   Class Preferences

My [INSERT GRADE LEVEL] students seem particularly interested in [INSERT INTERESTS]. Explain five ways to incorporate these interests into a lesson about [INSERT CONCEPT] to keep their attention.

ChatGPT Prompt Idea #158   Student Slang

How can I incorporate silly [INSERT GRADE LEVEL] slang into an engaging lesson about [INSERT MATH CONCEPT]?

ChatGPT Prompt Idea #159   Common Mistakes

What are some common mistakes that [INSERT GRADE LEVEL] students tend to make regarding [INSERT MATH CONCEPT]? How can I illustrate these mistakes in a way that helps the students avoid them?

ChatGPT Prompt Idea #160   Concept Differences & Celebrity Guests

Explain the difference between [INSERT MATH CONCEPT] and [INSERT MATH CONCEPT] as though you are [INSERT CELEBRITY]. The response should be simple enough for [INSERT GRADE LEVEL] students to understand and engaging enough to hold their attention.

ChatGPT Prompt Idea #161   Create a Song

Create a song to help [INSERT GRADE LEVEL] students remember [INSERT FORMULA OR MATH CONCEPT]. Make it to the tune of [INSERT POPULAR SONG].

ChatGPT Prompt Idea #162   Unlimited Examples

Please provide ten more examples to illustrate the concept of [INSERT CONCEPT] based on these ones:

[INSERT OTHER EXAMPLES]

You can then follow up with prompts asking for easier or more challenging examples.

ChatGPT Prompt Idea #163 . Discussion Questions

What are some open-ended discussion questions I could ask my [INSERT GRADE LEVEL AND CLASS] students to get them to fully understand the concept of [INSERT MATH CONCEPT]? Include grade-appropriate hints for each question.

ChatGPT Prompt Idea #164 . Building Confidence

What are some ways I can help build my [INSERT GRADE LEVEL] students' confidence with [INSERT MATH CONCEPT]? They seem to do well with [INSERT TRAIT] but struggle with [INSERT TRAIT(S)].

ChatGPT Prompt Idea #165 . Gamification

Create a game for [INSERT GRADE LEVEL] students to teach them about [INSERT MATH CONCEPT] based on the popular quiz show. Provide all instructions, questions, and answers.

ChatGPT Prompt Idea #166   Math Jokes

Generate 20 school-appropriate math jokes about [INSERT MATH CONCEPT] to be used with the warm-up in a [INSERT GRADE LEVEL] class.

ChatGPT Prompt Idea #167   Instant Grading

[INSERT IMAGE OF STUDENT'S WORK]

Analyze this [INSERT GRADE LEVEL] student's work and provide detailed feedback for every incorrect solution. Explain how to arrive at the correct solution step by step. Also, make suggestions for further practice.

ChatGPT Prompt Idea #168   Different Order, Same Solution

Show the different ways you can solve the same problem to arrive at the same solution:

[INSERT PROBLEM]

ChatGPT Prompt Idea #169   Math Scavenger Hunt

Create a list of math problems based on real-life scenarios for a scavenger hunt around the classroom. Each solution should lead to a new clue.

ChatGPT Prompt Idea #170 — Comprehensive Lesson Plan

I need to create a comprehensive lesson plan for a [INSERT TIMEFRAME] unit about [INSERT TOPIC] for my [INSERT GRADE LEVEL AND SUBJECT COURSE]. The lesson plan should include an overview of every [DAY/WEEK] as well as any supplies and prep that will need to be done ahead of time. Please also include differentiated learning paths for students who need fewer and more challenges.

Keep in mind the difficulties that students often have with learning math concepts.

CHAPTER 8
ENHANCING LITERACY AND READING ENJOYMENT WITH CHATGPT

"The mind is not a vessel to be filled but a fire to be kindled."
—Plutarch, Greek Philosopher

READING QUITE LITERALLY CHANGES the way our brain works by forcing it to expand and create new connections. How wild is that? As our neurons fire, we work out the parts of our brain responsible for our senses, stress relief, and memory. Some studies have even found that reading can reduce our chances of developing Alzheimer's or dementia. That's happy news for those of us who still remember the "Hooked on Phonics" days. Unfortunately, I don't have the happiest news to share when it comes to the state of reading in the US.

Ready for some rapid-fire statistics? Brace yourself. They're even more intense than the math statistics I shared in the last chapter.

HOW MUCH IS AMERICA REALLY READING?

These statistics come from Golden Steps and Scholastic:

- Fewer than 1/3 of Americans read an eBook for all of 2023
- The average reading level of American adults is 8th grade
- Approximately 1/3 of 4th graders and 8th graders are reading at or above their grade level
- 1/5 of Americans read below a 5th-grade level
- Nearly 1/2 of Americans read at or below a basic level
- The US barely makes it in the top 20 countries in the world in terms of reading proficiency
- The average American reads 12 books per year
- The majority of adults never visited a bookstore between 2018-2023
- Over half of low-income families have no books inside their home

SCREENS HAVE ENTERED THE CHAT

The statistics in the last sections start to make a little more sense when you learn these additional facts (Golden Steps):

- 50% of Americans own a tablet or eReader
- Nearly 1/2 of children get their own tablet device before elementary school ends
- 8-12 years olds average 4.5 hours of screen time per day
- Teens average nine hours of screen time per day
- Increased screen time is associated with disturbances in sleep, mood, and cognition

READING MAKES A DIFFERENCE

It's not all doom and gloom, though!

- A little over half of children who read eBooks prefer print books to eBooks
- Nearly all children who read eBooks read more than they traditionally would otherwise
- Children who read at home are much more likely to be successful in their studies
- Children whose loved ones read to them get exposed to nearly 2 million more words per year than their classmates
- Children who read for fun tend to have better health outcomes and achieve more

CHATGPT TACKLES LITERACY

Our favorite chatbot might just be one of the most voracious readers on this planet. Who else do you know that's quite literally read thousands upon thousands of books and can access their knowledge about any single one of them at the drop of a hat? Pretty impressive, if you ask me.

The model can be particularly useful with the following:

Personalized Reading Experiences

- **Tailored Reading Lists:** ChatGPT can create instant reading recommendations tailored to reading levels and genre preferences.
- **Interactive Storytelling:** ChatGPT can create dynamic stories from scratch or build off your ideas.

- **Reading Comprehension Assistance:** ChatGPT can serve as a reading comprehension tool by generating unlimited passages and practice opportunities.

Writing with AI

- **Creative Writing Prompts:** ChatGPT can share non-traditional ideas and prompts for student writing.
- **Grammar and Style Improvement:** The model can be trained to provide feedback related to writing and grammar.
- **Collaborative Writing Projects:** ChatGPT can promote group work between writers and readers of different strengths and levels.

Critical Thinking and Analysis

- **Analyzing Texts with AI:** ChatGPT can perform literary analysis across texts of varying difficulties.
- **Debate Preparation:** ChatGPT can aid in research and argument formulation.
- **Exploring Themes and Motifs:** ChatGPT can offer guided exploration of literary themes or devices, like foreshadowing, hyperbole, allegory, and allusion.

ChatGPT Prompt Idea #171   Summer Reading List

Create a list of mildly challenging books for students to read over the summer before entering [INSERT GRADE LEVEL]. The books should focus on the central theme of [INSERT THEME].

After narrowing down the list, you can then follow up with the following prompts:

Create a reading packet to go along with [INSERT BOOK TITLE]. The reading packet should ask open-ended and closed-ended questions along with other activities regarding the book to demonstrate reading mastery.

I need to write a letter to my [INSERT GRADE LEVEL] students' parents to tell them about the summer reading homework their students will need to complete. The letter should include my excitement for the upcoming course and explicitly state the requirements of the summer assignment. These are the requirements: [INSERT REQUIREMENTS].

ChatGPT Prompt Idea #172 . Unlimited Creative Writing Prompts

Generate a list of creative writing prompts for [INSERT GRADE LEVEL] students taking [INSERT CLASS]. The prompts should be age-appropriate and focus on [INSERT THEME]. The prompts should be mildly challenging.

ChatGPT Prompt Idea #173 . Common Grammar and Syntax Mistakes

What are the most common grammar and syntax mistakes that [INSERT GRADE LEVEL] students tend to make in their writing? Create a comprehensive guide I can pass out at the beginning of the quarter.

ChatGPT Prompt Idea #174 . Reading Comprehension Assistant

Create a list of multiple choice questions to assess reading comprehension based on the following text: [PASTE THE TEXT]

ChatGPT Prompt Idea #175 — Grammar Assistant

Provide feedback on the grammar of the following text written by a [INSERT GRADE LEVEL] student, pointing out strengths and areas for improvement: [PASTE TEXT].

ChatGPT Prompt Idea #176 — Persuasion Techniques

Provide an example of a persuasive argument for a class of [INSERT GRADE LEVEL] to analyze.

ChatGPT Prompt Idea #177 — Debate Topics

Create a list of age-appropriate and school-appropriate debate topics based on [INSERT BOOK] for a class of [INSERT GRADE LEVEL]. Include some basic ideas along with some more out-of-the-box ideas.

ChatGPT Prompt Idea #178 — Themes and Motif Application

How might the themes and motifs of [INSERT BOOK] apply to [INSERT GRADE LEVEL] students now and in the future?

ChatGPT Prompt Idea #179 — Generate Spelling Tests

Create a mildly challenging spelling word list for a [INSERT GRADE LEVEL] class.

You can add extra qualifiers about the theme of the words, the number of letters the words should be, or which letter each word should start with.

ChatGPT Prompt Idea #180 . Pop Quiz Generator

Create a short pop quiz based on [INSERT BOOK TITLE OR READING CONCEPT]. The pop quiz should be age-appropriate for [INSERT GRADE LEVEL]. It should consist of 5 easy to moderately challenging questions and two additional challenging questions for extra credit.

ChatGPT Prompt Idea #181 . Style Metamorphosis

Change the style of the following text to [INSERT OTHER STYLE] instead: [PASTE TEXT]

ChatGPT Prompt Idea #182 . Celebrity Guest

Pretend you are [INSERT CELEBRITY]. Provide a summary of [INSERT BOOK TITLE] for a group of [INSERT GRADE LEVEL] students.

ChatGPT Prompt Idea #183 . Remix

What are three alternative endings to [INSERT BOOK TITLE] that might have been interesting?

You might consider following up with either of these prompts:

Based on these alternative endings, create a list of open-ended questions for [INSERT GRADE LEVEL] students that would spark an interesting discussion

Based on these alternative endings, what are some engaging creative writing prompts I could assign a class of [INSERT GRADE LEVEL] students?

ChatGPT Prompt Idea #184 — Passage Generation

Generate a passage about [INSERT TOPIC] that would help [INSERT GRADE LEVEL] students practice their reading comprehension [OR OTHER READING SKILL]. Include a short quiz at the end of the passage to demonstrate understanding. The passage should be [INSERT IDEAL LENGTH], and the quiz should be [INSERT IDEAL LENGTH].

ChatGPT Prompt Idea #185 — What's the Difference?

Explain the difference between [INSERT LITERARY DEVICE] and [INSERT ANOTHER LITERARY DEVICE] in a way that's easy for [INSERT GRADE LEVEL] students to understand and relate to.

ChatGPT Prompt Idea #186 . Poem Generation

Generate a poem about [INSERT TOPIC] in the style of [INSERT FAMOUS POET OR CELEBRITY] that would engage a class of [INSERT GRADE LEVEL] students.

ChatGPT Prompt Idea #187 . Create the Perfect Library

What are some books every [INSERT GRADE LEVEL] teacher needs in their classroom? Offer an explanation for every book.

ChatGPT Prompt Idea #188 . Grammar Practice

Provide a sentence with grammatical errors. Have students identify and correct the errors.

ChatGPT Prompt Idea #189 . Practice Diagramming Sentences

Diagram the following sentences and explain your work: [PASTE TEXT]

ChatGPT Prompt Idea #190 . Cliffhangers

Provide the beginning of a story or a writing prompt, but make sure it ends on a cliffhanger. A class of [INSERT GRADE LEVEL] will finish the story.

ChatGPT Prompt Idea #191 Practice Inference

Share a short paragraph with implicit information. Ask students to make inferences about the characters, setting, or events.

ChatGPT Prompt Idea #192 If Book Characters Had Phones

What would a text exchange look like between [INSERT CHARACTER] and [INSERT OTHER CHARACTER] from [INSERT BOOK TITLE]?

ChatGPT Prompt Idea #193 Simplify the Text

Simplify the following text to help my class of [INSERT GRADE LEVEL] students understand it better: [PASTE TEXT]

ChatGPT Prompt Idea #194 Make It Shakespearean

Turn the following text into a sonnet written by Shakespeare: [PASTE TEXT]

ChatGPT Prompt Idea #195 Highlight Key Information

Underline the key parts of the following text and explain your reasoning: [PASTE TEXT]

ChatGPT Prompt Idea #196   Individualized Reading Lists

Create an individualized reading list for a [INSERT GRADE LEVEL] student who currently reads at the [INSERT GRADE NUMBER] level. The reading list should challenge them while also taking into consideration some of the books they have previously enjoyed.

Here is a list:

[INSERT BOOKS IN A BULLET POINT FORMAT]

ChatGPT Prompt Idea #197   Interview a Literary Character

You are [INSERT CHARACTER] from [INSERT BOOK], and our [INSERT GRADE LEVEL] would like to interview you. Answer all questions in character, and make sure your answers are school-appropriate.

ChatGPT Prompt Idea #198   Collaborate on a Story Together

Our [INSERT GRADE LEVEL] class would like to collaborate with you on a story. We will provide you with one sentence, and you will provide one sentence in return. We will continue taking turns until our story is complete. Confirm you understand the instructions.

Once the model confirms that they understand the activity, proceed by pasting your students' suggestions one by one into the chat.

ChatGPT Prompt Idea #199   Fill In the Blank Stories

Our [INSERT GRADE LEVEL] class would like to create a story together. We need you to ask us for a random number of nouns, adverbs, verbs, and adjectives. You will then use these random words to create a funny school-appropriate story.

The bot will then ask you for the information it needs. Respond with your students' chosen words, and it will instantly create your story.

ChatGPT Prompt Idea #200   Comprehensive Lesson Plan

I need to create a comprehensive lesson plan for a [INSERT TIMEFRAME] unit about [INSERT TOPIC] for my [INSERT GRADE LEVEL AND SUBJECT COURSE]. The lesson plan should include an overview of every [DAY/WEEK] as well as any supplies and prep that will need to be done ahead of time. Please also include differentiated learning paths for students who need fewer and more challenges.

Currently, most of my students are reading at the [INSERT GRADE NUMBER] level, with some reading at the [INSERT GRADE NUMBER] level

CHAPTER 9
MAKING SCIENCE COOL AGAIN WITH CHATGPT

"Everyone you will ever meet knows something you don't."
—Bill Nye The Science Guy, Beloved Scientist

IT'S ALWAYS a good day when Bill Nye or the Magic School Bus visits a science classroom because the students are engaged, and we get 30 minutes to catch up on our planning for the week or tie up other loose ends around the classroom. I've often heard students and educators alike joke about how much heavy lifting these two programs are doing for the science community, and it turns out they might be right.

The average elementary school class in the United States spends less than 20 minutes a day talking about science, and nearly three-quarters of elementary school teachers report feeling completely unprepared to teach the subject. What's more, less than one-fourth of high school students will be proficient in science by the time they graduate (National Academy of Sciences).

How can we change that?

Here's a start:
Virtual Labs and Simulations

- **Conducting Experiments Virtually:** We can use ChatGPT to simulate lab experiments by entertaining hypothetical scenarios like throwing a bowling ball and an egg off a building at the same time to learn about gravity.
- **Data Analysis and Interpretation:** We can teach students to analyze experimental data with AI as their personal lab assistant, which can teach them management, delegation, and teamwork skills that will serve them well in the future.

Enhancing Science Literacy

- **Scientific Writing Assistance:** ChatGPT can help improve our students' ability to write lab reports and research papers by giving them specific feedback or practice opportunities and examples.
- **Debunking Myths with AI:** We can use ChatGPT to challenge common scientific misconceptions, like the Tree Octopus fellow I mentioned earlier.
- **Real-World Problem Solving:** We can help our students apply scientific knowledge to solve global issues with ChatGPT as a source of multiple viewpoints and considerations.

Collaborative Science Projects

- **Global Science Fairs:** We can help our students see the bigger picture and how science collectively affects the rest of the world.
- **Sustainable Solutions Development:** We can use ChatGPT to brainstorm eco-friendly projects for the classroom and at home.

ChatGPT Prompt Idea #201 Fun Facts

Share an interesting science fact related to [INSERT THEME] that [INSERT GRADE LEVEL] students would enjoy.

ChatGPT Prompt Idea #202 Debunking Myths

Choose ten common science myths to debunk as a warm-up activity for a [INSERT GRADE LEVEL AND SUBJECT] class.

ChatGPT Prompt Idea #203 Putting the Scientific Method Into Practice

Come up with five practical ways a [INSERT GRADE LEVEL] student could use the Scientific Method in their daily life.

ChatGPT Prompt Idea #204 Hands-on Learning

Design a hands-on experiment for [INSERT GRADE LEVEL] students to learn about the concept of [INSERT TOPIC]. Recommend any necessary safety precautions in clear detail.

ChatGPT Prompt Idea #205 Incorporating Technology

What are some ways I can incorporate more technology into my lessons about [INSERT TOPIC]?

ChatGPT Prompt Idea #206 — Ethical Implications

Generate some discussion questions for a [INSERT GRADE LEVEL AND SUBJECT] class about the ethics of certain inventions and technologies.

ChatGPT Prompt Idea #207 — Lab Activity

What are some lab activities that can help [INSERT GRADE LEVEL] students observe [INSERT PHENOMENON]?

ChatGPT Prompt Idea #208 — Hypothesis Lesson

Craft a lesson for [INSERT GRADE LEVEL] students on what makes a good scientific hypothesis.

ChatGPT Prompt Idea #209 — Science Vocabulary

Create a list of all the vocabulary a class of [INSERT GRADE LEVEL] students should know on the topic of [INSERT TOPIC]

ChatGPT Prompt Idea #210 — Experiment Using Available Resources

What's a good science experiment for [INSERT GRADE LEVEL] students using some or all of the available classroom resources: [INSERT RESOURCES]

ChatGPT Prompt Idea #211 — Physics in Real Life

What are some practical applications of physics that [INSERT GRADE LEVEL] would find entertaining?

ChatGPT Prompt Idea #212 — Peer-Reviewed Research

How can I explain the importance of peer-reviewed research in a way that would make sense to [INSERT GRADE LEVEL] students?

ChatGPT Prompt Idea #213 — Simulations

Suggest some online simulations that [INSERT GRADE LEVEL] students could use to learn about [INSERT TOPIC]

ChatGPT Prompt Idea #214 — Creative Safety Guide

Create a fun and engaging yet practical safety guide for a [INSERT GRADE LEVEL] science class. Include science puns and jokes that are age-appropriate.

ChatGPT Prompt Idea #215 — Strategies to Memorize the Periodic Table

We are going to be hosting a contest for students to memorize the Periodic Table of Elements. What are some strategies that might help them remember the most element names?

ChatGPT Prompt Idea #216   Scientific Resources

What are some additional scientific resources my [INSERT GRADE LEVEL] students might find interesting for learning about [INSERT TOPIC]?

ChatGPT Prompt Idea #217   Analyzing the Hypothesis

I'd like you to analyze the data I've collected and tell me how it relates to my hypothesis.

[INSERT DATA]

ChatGPT Prompt Idea #218   Lab Report Tips

What are ten tips for [INSERT GRADE LEVEL] students to write better lab reports for a [INSERT SUBJECT] class?

ChatGPT Prompt Idea #219   Eco-Friendly Classroom

How can our [INSERT GRADE LEVEL AND SUBJECT] class be more eco-friendly? Help us brainstorm a list of 15 ideas.

ChatGPT Prompt Idea #220   Comprehensive Lesson Plan

I need to create a comprehensive lesson plan for a [INSERT TIMEFRAME] unit about [INSERT TOPIC] for my [INSERT GRADE LEVEL AND SUBJECT COURSE]. The lesson plan should include an overview of every [DAY/WEEK] as well as any supplies and prep that will need to be done ahead of time. Please also include differentiated learning paths for students who need fewer and more challenges.

The lesson plan should include some lab-based activities, along with any important safety measures that we should consider.

CHAPTER 10
BRINGING THE PAST TO LIFE WITH CHATGPT

"A teacher affects eternity; He can never tell where his influence stops."

—Henry Brooks Adams, American Historian

COMPETENCY ACROSS HISTORY, geography, and civics has been plummeting, with less than 1/5 of 8th graders scoring proficient or above in U.S. history in the past four years (Wexler). If we zoom out further, nearly half of students scored below the basic level of knowledge in U.S. history (Fensterwald)...which begs the question, what are we doing?

Studies show most districts tend to focus more on reading and math due to the growing emphasis on state testing, leaving history dead last on the priority list (Hutton et al.). And what does that leave us with? Students who can't point out a single country on a map of the world, students who don't know the outcome of the Civil War, and students who have no clue when the Constitution was ratified.

I don't mean to be a downer, but hiding away from these facts and statistics does our students no good. If we are to face this frightening trend, we must do so head-on. Looking to administration for support sounds nice in theory, but it hasn't necessarily worked out much in practice.

If we only have 5 minutes, 20 minutes, 40 minutes—whatever it may be—to teach history so that it doesn't one day repeat itself, let's make it count.

ChatGPT isn't living or breathing, but it is certainly the greatest time capsule the world has ever seen. Consider these three use cases:

Interactive Historical Narratives

- **Time Travel Dialogues:** We can engage our students with historical figures through ChatGPT conversations that work like interviews.
- **Analyzing Primary Sources:** We can use ChatGPT to help students understand the importance and context surrounding certain historical documents.
- **Virtual Field Trips:** We can plan AI-driven field trips through virtual archives using ChatGPT's recommendations.

Connecting Past and Present

- **Thematic Studies:** We can use ChatGPT to draw parallels between historical events and current issues and encourage our students to apply historical lessons to modern challenges.

- **Cultural Heritage Projects:** We can promote the understanding and appreciation of diverse cultures with ChatGPT's help.

Critical Thinking in History

- **Debating Historical Perspectives:** ChatGPT can tap into its knowledge base to suggest interesting debate topics outside the usual repertoire.
- **Bias and Source Analysis:** We can teach our students to evaluate historical narratives and sources more critically.
- **Constructing Historical Arguments:** We can use ChatGPT to help our students craft and better support their historical arguments.

ChatGPT Prompt Idea #221   Reading Historical Diaries

Imagine you were alive during [INSERT TIME PERIOD]. Write a diary entry about what it was like.

ChatGPT Prompt Idea #222   Creating Flashcards

Create a set of flashcards for a unit about [INSERT TOPIC] for [INSERT GRADE LEVEL] students.

ChatGPT Prompt Idea #223   Time Traveling

Pretend you time traveled back to [INSERT YEAR] with a [INSERT MODERN DAY INVENTION]. What might the people around you say about this invention?

ChatGPT Prompt Idea #224 Reimagining History

How might history have changed if [INSERT HISTORICAL FIGURE] never existed?

ChatGPT Prompt Idea #225 Rhyme Time

Create rhymes to help a class of [INSERT GRADE LEVEL] students memorize [INSERT HISTORY CONCEPT].

ChatGPT Prompt Idea #226 Provide Short Summaries

Generate a concise summary of the events of [INSERT WAR] using a bullet point format. Note key dates and historical figures.

ChatGPT Prompt Idea #227 Virtual Field Trip

Craft a virtual field trip for a group of [INSERT GRADE LEVEL AND SUBJECT] students studying about [INSERT TIME PERIOD].

ChatGPT Prompt Idea #228 Teaching Students Perspective-Taking

Create a writing prompt that asks [INSERT GRADE LEVEL] students to consider [INSERT MAJOR HISTORICAL EVENT] from the perspective of each of the parties involved.

ChatGPT Prompt Idea #229   History and Ethics

Come up with a list of school-appropriate topics that [INSERT GRADE LEVEL] students can debate regarding history and ethics.

ChatGPT Prompt Idea #230   Analyzing Historical Documents

Act as a [INSERT GRADE LEVEL] tutor. Explain the importance of [INSERT HISTORICAL DOCUMENT] as it relates to present day. Point out specific terminology and explain the historical context of the document.

ChatGPT Prompt Idea #231   Constitution Assignment

Create an outline for an assignment that asks [INSERT GRADE LEVEL] students to craft their own constitution for their family.

ChatGPT Prompt Idea #232   Create a Historical Slideshow Presentation

I need to create a slideshow presentation about [INSERT HISTORY TOPIC]. Map out a 15-minute presentation for my class of [INSERT GRADE LEVEL] students.

ChatGPT Prompt Idea #233   Historical Context

Craft an assignment for a [INSERT GRADE LEVEL AND SUBJECT] class that demonstrates why historical context is so important regarding past places, people, and events.

ChatGPT Prompt Idea #234 — Bias in History

Craft an assignment for a [INSERT GRADE LEVEL AND SUBJECT] class that helps students understand how bias occurs in history, how it can be dangerous, and how to prevent it.

ChatGPT Prompt Idea #235 — Culture Time Capsule

Assign a group of [INSERT #, GRADE LEVEL, AND SUBJECT] students different cultures that they will have to consider when creating a time capsule. Then create the assignment instructions. It should be a written assignment no more than [INSERT PAGE LENGTH] and use [INSERT FORMAT].

ChatGPT Prompt Idea #236 — Gratitude Journal

Assign students a time period in history. Then ask them to pick one person they are grateful for from that time period. Students should write this person a letter as a journal assignment. Include a rubric that should assess students on the following: [INSERT QUALITIES]

ChatGPT Prompt Idea #237 — History Comes to Life

We are going to celebrate the end of the quarter by having our [INSERT GRADE LEVEL AND SUBJECT] class dress up as historical figures. Provide a list of 20 ideas for our students to dress up as, along with simple costume ideas.

ChatGPT Prompt Idea #238 . Greek Olympics Challenge

Our entire [INSERT GRADE LEVEL] class will be competing in our own Greek Olympics. Come up with 10 potential challenges that our students could do throughout the school building related to the original Olympic games.

ChatGPT Prompt Idea #239 . Hieroglyphs Activity

We are learning about hieroglyphs. How can we incorporate this concept into a fun activity for [INSERT GRADE LEVEL] students?

ChatGPT Prompt Idea #240 . Bite-Size History

I would like to start introducing mini 5-minute history lessons into my [INSERT GRADE LEVEL] class to help my students learn more about [INSERT TOPIC]. Give me a list of ten ideas.

ChatGPT Prompt Idea #241 . Historical Playlist

Help me craft a musical playlist that showcases history going as far back as the classical times all the way up to the 2000s. Make sure it is school-appropriate.

ChatGPT Prompt Idea #242   Other History Songs

What are some famous songs about historical concepts? Make sure they are kid-friendly.

ChatGPT Prompt Idea #243   Time Traveling Celebrity

If [INSERT CELEBRITY] traveled back in time to [INSERT TIME PERIOD], what would they do? Create an entertaining story for a group of [INSERT GRADE LEVEL AND SUBJECT] students.

ChatGPT Prompt Idea #244   Memorizing Country Names

What are some mnemonic devices or other memory devices that my [INSERT GRADE LEVEL AND SUBJECT] students could use to memorize other country names?

ChatGPT Prompt Idea #245   Comprehensive Lesson Plan

I need to create a comprehensive lesson plan for a [INSERT TIMEFRAME] unit about [INSERT TOPIC] for my [INSERT GRADE LEVEL AND SUBJECT COURSE]. The lesson plan should include an overview of every [DAY/WEEK] as well as any supplies and prep that will need to be done ahead of time. Please also include differentiated learning paths for students who need fewer and more challenges.

The lesson plan should include how history connects to the present day.

CHAPTER 11
DON'T FORGET ABOUT THE ELECTIVES

"In learning, you will teach, and in teaching, you will learn."
— Phil Collins, Musician

BY NOW, I hope I've made it clear that ChatGPT's limitations are what you make them. You can prompt to your heart's content for any subject you could possibly think of, even outside of the core curriculum. This book can't possibly offer an exhaustive list, so I've taken a handful of the most popular electives across K-12 classrooms and offered up a few additional prompts to get the most out of your ChatGPT subscription.

Author's Note: If you have more prompt ideas you'd like to share with me, you can reach me at admin@halcyontimespress.com. They might just end up in the next version of this book!

ART CLASS

ChatGPT Prompt Idea #246 , Color Theory

I would like to teach [INSERT GRADE LEVEL] students about color theory. Provide five humorous examples of color theory that they can probably relate to.

ChatGPT Prompt Idea #247   Beginner-Friendly

What is a beginner-friendly art project idea that would teach [INSERT GRADE LEVEL] students about [INSERT CONCEPT]?

ChatGPT Prompt Idea #248   Art History

Create a list of art project ideas that could illustrate several different moments in art history together. The project should be developmentally appropriate for [INSERT GRADE LEVEL] students.

ChatGPT Prompt Idea #249   Limited Resources

Our art class is running low on resources. This is what we have available:

[INSERT LIST OF RESOURCES]

What are some project ideas for [INSERT GRADE LEVEL] students?

ChatGPT Prompt Idea #250   Specific Artist

What are some art project ideas for something inspired by [INSERT ARTIST]? Please include project ideas of varying degrees of difficulty for [INSERT GRADE LEVEL] students.

PHYSICAL EDUCATION

ChatGPT Prompt Idea #251 , Motor Skills

Give me five suggestions for games that will improve motor skills in [INSERT GRADE LEVEL] students.

ChatGPT Prompt Idea #252 , Locker Room Rules

I need to create a poster for our [INSERT GRADE LEVELS] gym locker room that covers the rules. Two strict rules are no photography and no bullying. What else should I include?

ChatGPT Prompt Idea #253 , Teamwork

What are some cardio-based activities that a class of [INSERT GRADE LEVEL] students could do together during P.E. class to build their teamwork skills?

ChatGPT Prompt Idea #254 , Alternative Assignment

One of my gym class students has a doctor's note and cannot participate in any physical activity. Help me come up with an alternative assignment that they can complete to still earn course credit.

ChatGPT Prompt Idea #255   Personal Fitness Plan

This week, our [INSERT GRADE LEVEL] P.E. class will be meeting in the weightlifting room. I'd like to develop a template to help each student come up with their own goals for the week and create their own fitness plan.

MUSIC CLASSES

ChatGPT Prompt Idea #256   Range Suggestions

What are some good song ideas for an [INSERT GRADE LEVEL] [INSERT INSTRUMENT] solo song in the [INSERT RANGE]?

ChatGPT Prompt Idea #257   Sightreading Tips

What are some sightreading tips for [INSERT GRADE LEVEL] students?

ChatGPT Prompt Idea #258   Vocal Exercises

What are some fun vocal exercises that a class of [INSERT GRADE LEVEL] students might enjoy? They should be light-hearted and humorous while remaining school-appropriate.

ChatGPT Prompt Idea #259   Rhythm Exercises

My students are struggling with rhythm. What are some unique ways we can work on this school together?

ChatGPT Prompt Idea #260 Unusual Instruments

Offer suggestions for common classroom supplies our class can safely use as instruments.

LANGUAGE CLASSES

ChatGPT Prompt Idea #261 Cultural Immersion with Names

You are a seasoned [INSERT GRADE LEVEL] [INSERT LANGUAGE] teacher. Brainstorm a list of culturally appropriate names for the students to choose from. Make sure there are female, male, and unisex options.

ChatGPT Prompt Idea #262 First Day of Class

What are some fun activities for [INSERT GRADE LEVEL] students who are starting a [INSERT LANGUAGE] class for the first time?

ChatGPT Prompt Idea #263 Guess That Song

Translate these popular song lyrics from English to [INSERT OTHER LANGUAGE]:

[PASTE ORIGINAL LYRICS]

Then, have the students guess the song.

ChatGPT Prompt Idea #264 . Cultural Imersion Menu

Put together a menu for a [INSERT LANGUAGE] class of [INSERT GRADE LEVEL] students to enjoy together during a day of cultural immersion. Include specific recipes and ingredients.

ChatGPT Prompt Idea #265 . Folklore and Myths

What are three common folk tales and myths that a [INSERT GRADE LEVEL] class would enjoy that pay homage to [INSERT LANGUAGE] roots? Explain how it relates to their culture.

ChatGPT Prompt Idea #266 . Celebration

Choose a holiday that [INSERT GRADE LEVEL AND CLASS] students could celebrate at school together. Give suggestions for activities that are culturally relevant.

ChatGPT Prompt Idea #267 . Fun Facts

Name three fun facts that [INSERT GRADE LEVEL] students might not know about [INSERT PLACE OR COUNTRY].

ChatGPT Prompt Idea #268 . Translation Project

What is a fun non-traditional translation project that could help [INSERT GRADE LEVEL] students practice their [INSERT LANGUAGE] skills?

ChatGPT Prompt Idea #269   Movie Scene Re-enactment

I would like my [INSERT GRADE LEVEL] students to pick a well-known movie scene of their choice in English, translate it to [INSERT LANGUAGE], and then re-enact it with a partner. What are some famous school-appropriate movie scenes?

ChatGPT Prompt Idea #270   Flag Project

What ideas do you have for a project that our [INSERT GRADE LEVEL AND LANGUAGE] class can do to learn about other country flags and pay respect to their cultures?

CHAPTER 12
ENCOURAGING CREATIVE PROJECTS AND CRITICAL THINKING

"Instruction does much, but encouragement everything."
—Johann Wolfgang von Goethe, Polymath and Writer

CREATIVITY MIGHT BE the last thing you think of when you hear the words Artificial Intelligence, but have you ever stopped to ask yourself why that might be the case? Are photographers any less talented than painters? Are graphic designers any less talented than sketch artists? Are writers who use a laptop any less talented than those who tap away at a typewriter or scrawl into a tiny notebook with a blue ballpoint pen?

Technology has only allowed us to be more creative by untying us from the shackles of obligation. Writers, artists, and musicians no longer need to travel miles to a library to conduct research; we have access to a world of experiences and stories at our fingertips. We no longer need to fear writer's block when new ideas and new inspiration are only a few clicks away.

The prompts included in this book don't even account for 1% of ChatGPT's capabilities. *Let that sink in*. For every 300 prompts, there are 3,000+ opportunities to add parameters and tweak the conversation to produce an entirely different output.

Make these use cases your own!

Designing Project-Based Learning

- Planning and executing interdisciplinary projects
- Using ChatGPT for research and information gathering
- Making presentations more exciting with ChatGPT's recommendations
- Highlighting the Socratic Method and other discussion-based learning opportunities
- Encouraging students to take the lead with their learning

Encouraging Critical Thinking and Deeper Analysis

- Developing critical questioning techniques
- Opening debates and discussions with ChatGPT
- Analyzing biases and new perspectives
- Encouraging evidence-based reasoning
- Promoting digital literacy and media critique

Amplifying Creative Expression and Unlimited Brainstorming Opportunities

- Inspiring creative writing and storytelling
- Encouraging STEM projects
- Demonstrating entrepreneurial thinking
- Making resources stretch further or introducing new ones
- Showcasing student projects and case studies

ChatGPT Prompt Idea #271   Interdisciplinary Projects

What are some interdisciplinary project ideas for a class of [INSERT GRADE LEVEL] students learning about [INSERT TOPIC]? Provide ideas for solo work and group work of varying degrees of difficulty.

ChatGPT Prompt Idea #272   Research

What are three good resources to use for researching [INSERT TOPIC]?

ChatGPT Prompt Idea #273   Digital Literacy

What are five core tenets of digital literacy that every [INSERT GRADE LEVEL] student should know?

ChatGPT Prompt Idea #274   Debate Partner

Act as a [INSERT GRADE LEVEL] debate partner. You and I will take turns debating about [INSERT TOPIC]. We will each get [INSERT #] turns to make our point. Then, you will transform into a debate coach and analyze our conversation.

ChatGPT Prompt Idea #275   Opposing View

Analyze this paragraph and offer three opposing viewpoints that the author might not be considering. [INSERT PARAGRAPH]

ChatGPT Prompt Idea #276 — Business Ventures

Give [INSERT GRADE LEVEL] students three unique business ideas for a mock entrepreneurship project. Outline the six-week project and expectations.

ChatGPT Prompt Idea #277 — STEM Opportunities

How can I get my [INSERT GRADE LEVEL] class more excited about STEM and infuse it into our daily lessons in a way that is natural?

ChatGPT Prompt Idea #278 — Ask the Experts

What would an interdisciplinary team of experts have to say about the following argument? [INSERT ARGUMENT]

ChatGPT Prompt Idea #279 — Headlines

Demonstrate how one news headline can take on many different meanings just by changing a few words.

ChatGPT Prompt Idea #280 — Checking for Bias

Please analyze my work and let me know if there are any potential biases. Offer specific suggestions and examples to fix them.

ChatGPT Prompt Idea #281   Case Studies

Tell me about some successful student entrepreneurs.

ChatGPT Prompt Idea #282   Deduction vs. Induction

Explain the difference between deduction and induction to a class of [INSERT GRADE LEVEL] students using examples that they'd find humorous.

ChatGPT Prompt Idea #283   Types of Questions

Brainstorm a list of open, closed, clarifying, speculative, dichotomous, and leading questions about [INSERT TOPIC] that would help a class of [INSERT GRADE LEVEL] students work on their critical and creative thinking skills.

ChatGPT Prompt Idea #284   Boolean Phrases

I am trying to research [INSERT TOPIC]. What are some good Boolean phrases or keywords I should use?

ChatGPT Prompt Idea #285   Evaluate the Claim

Evaluate the following claim using only facts: [INSERT CLAIM]

CHAPTER 13
TRANSFORMING ASSESSMENT WITH CHATGPT

"Knowledge will bring you the opportunity to make a difference."
—Claire Fagin, American Nurse and Educator

THE OTHER CHAPTERS have held some pretty harsh truths about the state of education in this country, but are you ready for some good news? *There is some light on the horizon.*

As more states push for through-year assessment models, less emphasis is being placed on beginning-of-year or end-of-year milestones (Pearce). Instead, students and teachers can close achievement gaps together by pacing the curriculum more appropriately. As of the beginning of the year, 13 states are already on board, including Florida, Texas, Louisiana, and Montana.

Many states are also moving away from multiple-choice style tests and instead opting for performance-based tests, which tend to allow more open-ended answers. Colorado, Kentucky, Massachusetts, and Nebraska are leading the charge on these.

Lastly, high schools across the country are opting out of exit examinations as part of their graduation requirements, allowing students to focus more on holistic college prep instead. Washington, for example, now offers soon-to-be graduates eight different pathways to demonstrate their college and career readiness.

As for how ChatGPT is changing the game:

Dynamic Assessments: Creating Adaptive Tests and Quizzes

- Designing personalized quizzes
- Tweaking formative and summative assessments
- Using ChatGPT to develop oral assessments
- Developing problem-solving challenges
- Offering Pre-tests and Post-tests
- Engaging students in self-assessment

Instant and Constructive Feedback: Using AI to Get Back to Students Much Quicker

- Balancing positive reinforcement with constructive critique
- Automating routine feedback, saving time for personalized guidance
- Creating templates for quicker feedback
- Giving students follow-up prompts to keep improving

Long-Term Progress Tracking: Monitoring Growth Over Time

- Setting up student learning portfolios
- Analyzing trends and identifying areas for improvement
- Customizing learning goals and milestones
- Leading parent-teacher discussions with data
- Preparing students for standardized tests with predictive analytics

ChatGPT Prompt Idea #286 — Assessment Revision

I want to make sure my assessments are fair and accurate. Please provide constructive feedback on my [INSERT GRADE LEVEL AND SUBJECT] assessment:

[INSERT ASSESSMENT]

ChatGPT Prompt Idea #287 — Positive Report Card Feedback

I need to draft report card comments for my [INSERT GRADE LEVEL AND SUBJECT] students. I'd like to conduct three comment frameworks.

1. For students who have shown a lot of improvement in our class
2. For students who are struggling and might need some extra support
3. For students whose parents I would like to talk to further about support needs

ChatGPT Prompt Idea #288 — Oral Assessment

I would like to create an oral assessment for my [INSERT GRADE LEVEL AND SUBJECT] class about [INSERT TOPIC]. There should be a mix of True/False, Fact-based questions, and Open-ended questions.

ChatGPT Prompt Idea #289   Alternative Assignments or Assessments

What are some alternate assignment ideas that can be used to assess the following skills: [INSERT SKILLS]

OR

What are some ways I can assess my students' knowledge on [INSERT TOPIC] without making them take a quiz or test?

ChatGPT Prompt Idea #290   Data Collection

Create a list of 5 types of qualitative and quantitative data that I can collect to monitor the progress of my students.

ChatGPT Prompt Idea #291   Identifying Knowledge Gaps

What are some questions I can ask my class to identify strengths and gaps in their learning?

ChatGPT Prompt Idea #292   Pre-Test

I'd like to develop an ungraded pre-test for my [INSERT GRADE LEVEL AND SUBJECT] students about [INSERT TOPIC] to give them a preview of the upcoming lesson.

ChatGPT Prompt Idea #293   ACT and SAT Prep

What are some warm-up activities that my [INSERT GRADE LEVEL AND SUBJECT] class can start doing to help them prepare for the [ACT/SAT]?

ChatGPT Prompt Idea #294 . Career Inventory

Suggest some career inventory tests that might help my [INSERT GRADE LEVEL] students figure out what they want to do with their future.

ChatGPT Prompt Idea #295 . College Readiness

Even though my students are only in the [INSERT #] grade, what are some ways we can start talking about college readiness?

ChatGPT Prompt Idea #296 . Self-Assessment

I would like to create a self-assessment for all my [INSERT GRADE LEVEL AND SUBJECT] students to submit with all their assignments this quarter. Include guidelines for being honest and taking time for intentional self-reflection.

ChatGPT Prompt Idea #297 . Sandwich Feedback

Please illustrate the sandwich feedback method using examples that would appeal to a [INSERT GRADE LEVEL] class.

ChatGPT Prompt Idea #298   Problem-Solving Practice

Our [INSERT GRADE LEVEL] class needs to work on our critical and creative thinking skills. Give us a silly problem that we can collaborate on to solve. We will then share the solutions we come up with. Analyze them and provide feedback on the points we may not have considered.

ChatGPT Prompt Idea #299   Formative Assessment Plan

I need to create a formative assessment plan for my [INSERT GRADE LEVEL] class that helps my students reach the following benchmarks: [INSERT BENCHMARKS]. Include strategies for the students and teacher to be successful. Ask any clarifying questions you may need to get the full picture before creating the plan.

ChatGPT Prompt Idea #300   Summative Assessment Plan

I need to create a summative assessment plan for my [INSERT GRADE LEVEL] class that helps my students reach the following benchmarks: [INSERT BENCHMARKS]. Include strategies for the students and teacher to be successful. Ask any clarifying questions you may need to get the full picture before creating the plan.

CHAPTER 14
PROFESSIONAL DEVELOPMENT AND LIFELONG LEARNING WITH CHATGPT

"Anyone who stops learning is old, whether at twenty or eighty. Anyone who keeps learning stays young."

—Henry Ford, American Industrialist and Businessman

YOU'RE in this field to help others, which is why you now spend 80 hours a week doing just that. But what about you? Think you can spare 10 minutes to dedicate to your own growth and learning? If it's permission you're looking for, you've got mine. And if that doesn't work, then I dare you to keep reading.

If there's one word I've come to despise over the years, it's "just." It usually flies under the radar in most conversations, but it pierces my eardrums every time I hear it. Unfortunately, it might start doing the same to yours because once you notice the pattern, you won't be able to UN-notice it. It usually sounds something like this:

Oh, I'm JUST a teacher.

Ouch! Did you feel it? How many of our colleagues walk around carrying that sentiment in their heads and their hearts, so much so that it casually slips out in every interaction with a stranger who's trying to make small talk?

Let's toss it. There's no "just" around here. There's simply:

I'm a teacher!!!

Anywho, ChatGPT isn't *just* for our students (Sorry, I couldn't resist).

Here's what's in it for us:

Self-Directed Learning: Using ChatGPT for Your Own Growth

- Accessing professional development resources
- Customizing learning paths for educators
- Engaging in AI-assisted reflective practice
- Staying updated with the latest educational technology trends
- Building a personal learning network

Collaborative Learning Communities: Creating Networks of Knowledge

- Joining online forums and discussion groups
- Organizing virtual conferences and workshops
- Sharing best practices and teaching resources
- Collaborating on curriculum development
- Participating in peer coaching and mentoring

Innovating Teaching Practices: Experimenting with New Methods

- Integrating emerging technologies in the classroom
- Developing interdisciplinary approaches
- Exploring global education initiatives
- Piloting novel assessment techniques
- Leading change in educational institutions

Career Expansion: Conducting Job Prep with AI

- Editing resumes or finding keywords to make them stand out

- Conducting mock job interviews
- Inquiring about different career paths
- Preparing for performance reviews, board meetings, and shadowing
- Drafting follow-up emails and messages

ChatGPT Prompt Idea #301   Professional Development Activities
What are some professional development activities that could help teachers become more comfortable with technology?

ChatGPT Prompt Idea #302   Interview Prep
Help me prepare for a job interview. I will be interviewing for the position of [INSERT TITLE]. I would consider my best skills [INSERT SKILLS], and I have [INSERT #] years of experience.

ChatGPT Prompt Idea #303   Teacher Collaboration
What are some ways that our school can be more proactive in sharing our teaching philosophy and fun activities for our students?

ChatGPT Prompt Idea #304   Resume Coach
Provide feedback on my resume as an expert resume coach and education recruiter. [INSERT RESUME]

ChatGPT Prompt Idea #305   Email Proofreader
Please proofread my email for grammar, spelling, and punctuation errors. Also, make suggestions related to the tone: [INSERT DRAFT]

ChatGPT Prompt Idea #306   Career Coach
I am a [INSERT JOB TITLE], and I would like to eventually

become a [INSERT DIFFERENT JOB TITLE]. Act as an expert career coach and help me create a plan to work towards that goal.

ChatGPT Prompt Idea #307 . Professional Development Resources
What are some professional development resources I might not have considered for teachers?

ChatGPT Prompt Idea #308 . Getting Published
I am a teacher who would eventually like to get published. What are the steps I would need to take to make that happen? Ask me any relevant questions to get a clearer picture of my goals.

ChatGPT Prompt Idea #309 . Side Hustle Ideas
My teacher's salary isn't cutting it. What are some ideas for side hustles I can do that don't take too much extra time or energy? Share the realistic earnings and pros and cons of each.

ChatGPT Prompt Idea #310 . Global Education Initiatives
What global education initiatives have there been over the last decade that I should know about?

ChatGPT Prompt Idea #311 . Recommendation Letter Templates
I need to create a recommendation letter template that I can use for my students who are applying to scholarships and colleges. The template should be personable and professional in nature and leave room for personalization.

ChatGPT Prompt Idea #312 . Professional Wardrobe
Where can I find new clothes as a [INSERT GENDER] teacher looking for [INSERT STYLE] [INSERT GARMENT TYPE]

ChatGPT Prompt Idea #313 • Preventing Burnout

What are some realistic strategies I can implement to help me prevent burnout as a teacher? Skip the cookie-cutter advice.

ChatGPT Prompt Idea #314 • Networking Events

What are some networking events or seminars I should consider attending as a [INSERT GRADE LEVEL] teacher?

ChatGPT Prompt Idea #315 • Collaborative Classes

How can my [INSERT GRADE LEVEL] class collaborate with [INSERT OTHER CLASS] to teach our students [INSERT LIST OF SKILLS]? Provide 5 project or activity ideas.

ChatGPT Prompt Idea #316 • Congratulating Colleagues

Draft a courteous, professional email to [INSERT TEACHER'S NAME] congratulating them on [INSERT ACCOMPLISHMENT].

ChatGPT Prompt Idea #317 • New Hobbies

I feel like I've been so busy teaching that I haven't been able to try any new hobbies lately. Give me 5 ideas for hobbies that are low-budget and don't take up too much time. Ask me questions to better understand my preferences and current skills.

ChatGPT Prompt Idea #318 • Meeting Agenda and Reminders

Create an agenda for our upcoming meeting about [INSERT TOPIC]. Also, include email reminders for a week prior and the day prior.

ChatGPT Prompt Idea #319 • Stress Relief

What are some good stress relief ideas for a teacher that don't involve breathing or yoga?

ChatGPT Prompt Idea #320 — Budgeting
I am a teacher who would like to learn more about budgeting. Can you help me craft a reasonable budget?

ChatGPT Prompt Idea #321 — Mixing It Up
What are some new teaching methods I could try to mix things up for a class of [INSERT GRADE LEVEL] students?

ChatGPT Prompt Idea #322 — Leading Initiatives
I would like to lead the initiative for [INSERT TOPIC] at my [INSERT TYPE] school. Help me come up with an idea to help get the school board, other teachers, and parents on board.

ChatGPT Prompt Idea #323 — Teacher Reflections
Act as an expert teaching coach and ask me questions to help me reflect on my teaching journey and teaching goals.

ChatGPT Prompt Idea #324 — Hosting Workshops
You are a digital marketing expert. I would like to host a workshop for [INSERT AUDIENCE] about [INSERT TOPIC]. Help me create a game plan that outlines everything from the pre-launch to the live event and post-launch.

ChatGPT Prompt Idea #325 — Assessment Trends
What are some assessment trends for [INSERT GRADE LEVEL] students over the past decade? Share relevant research and resources.

ChatGPT Prompt Idea #326   Making Teacher Friends
Act as a professional life coach. Help me come up with a plan to connect with the other teachers at my school.

ChatGPT Prompt Idea #327   Scheduling
I feel like my schedule has been all over the place, and I am struggling to balance work with my personal life. Help me create a more manageable schedule by asking me relevant questions related to my work obligations, personal obligations, and hobbies.

ChatGPT Prompt Idea #328   Managing Work Conflict
Act as a professional career coach and help me brainstorm solutions for the conflict I have been experiencing at work. Help me understand the conflict from multiple perspectives.
[INSERT CONTEXT]

ChatGPT Prompt Idea #329   Elevator Speech
You are a career coach. Help me create the perfect elevator speech that tells others what I do and how I can help them.

ChatGPT Prompt Idea #330   Social Media Content
You are a digital marketing expert. I would like to start creating social media content for the following platforms: [INSERT PLATFORMS]. Help me brainstorm a month of content for each platform as a teacher who isn't really sure where to start.

ChatGPT Prompt Idea #331   Finding a Mentor
I am a [INSERT GRADE LEVEL] teacher who is looking for a mentor to help me [INSERT CONTEXT]. How would I go about connecting with someone like that?

ChatGPT Prompt Idea #332 Date Night

I would like to plan more date nights with my partner. Here are some common interests we share: [INSERT LIST]. Give me some ideas for budget-friendly dates that two adults would enjoy.

ChatGPT Prompt Idea #333 Vacation Planning

You are an expert travel planner. I am a teacher who has summers off. I would like to plan a vacation for my family, which consists of [INSERT # OF PEOPLE AND AGES]. We are willing to travel [INSERT DISTANCE] and would prefer to stay under [INSERT BUDGET]. Help us plan the perfect vacation.

ChatGPT Prompt Idea #334 Crafting

I would like to get more into crafting, but I'm not sure where to start. What are some beginner-friendly crafts an adult might like?

ChatGPT Prompt Idea #335 Family Time

I'm a teacher who spends a lot of time planning and working. Help me come up with a list of ways I can be more intentional about family time with my [INSERT FAMILY MEMBERS AND AGES].

CHAPTER 15
MOST COMMON QUESTIONS TEACHERS HAVE ABOUT AI (AND YOUR ANSWERS)

"Education breeds confidence. Confidence breeds hope. Hope breeds peace."

—Confucius, Chinese Philosopher

WE'RE NEARING the end of our journey together, but we're not quite done yet. You won't find any more prompts from here on out, but what I have compiled is a list of the most common questions teachers have had about AI sourced from various forums. I have attempted to present all sides so you can form your own judgments about the material presented.

By far, the most asked question was as follows:

WHAT ARE YOUR BEST TRICKS FOR CATCHING AI-GENERATED SUBMISSIONS?

You might not like this answer, but there is no definitive foolproof way to catch AI-generated submissions, despite what some overly confident educators might be rattling off to Reddit and Quora (I'll explain this further in the next question). However, two strategies have seemed to gain the most traction:

1. Adding white "invisible" text to the instructions
2. Requiring outlines or first drafts to be written in class

The first strategy is to embed false instructions in "invisible ink" among the official assignment instructions. This could be a simple request like "mention apples and oranges" or something more complex like "analyze the following quote in your essay" and include a fake quote. Students who copy/paste the assignment instructions into any AI tool will then produce assignments with the false information.

The second strategy takes more effort on the educator's part by adding some extra "checks and balances" to the assignment. Outlines and first drafts must be written by hand in class and then pre-approved before students are allowed to work on the assignment outside of class.

Unfortunately, neither strategy is foolproof, as students in the first group may simply edit the extraneous information out before submission and then alert their classmates about the trick. Students in the second group can also continue to use AI outside the classroom despite their pre-work getting the human stamp of approval.

WHAT ARE SOME GOOD AI DETECTORS I CAN USE TO CHECK MY STUDENTS' WORK?

These are the most common AI detectors on the market right now:

- Copyleaks
- ZeroGPT
- Writer
- GPTZero
- Quillbot

However, their accuracy remains questionable, as I'll explain.

HOW ACCURATE ARE AI DETECTORS AND PLAGIARISM DETECTORS?

Several AI detectors like to market themselves as more than 99% credible and accurate, but these claims are **100% unfounded** and not based on any scientific measures whatsoever. I have personally witnessed colleagues write sample papers themselves, upload them to these tools, and have them result in a false-positive which mistakenly attributes the writing to AI. There is a chance these tools will continue to get more sophisticated as time goes on, but AI will do the same, and the AI models themselves have the advantage here.

I encourage you to try the AI detectors above and see the results for yourself. Copy/paste the same paragraph into five different detectors, and you'll get five different results.

WHAT'S THE DIFFERENCE BETWEEN AI-GENERATED AND AI-ASSISTED CONTENT?

Officially, AI-generated content means a machine model did 100% of the work, while AI-assisted content requires at least some percentage of human involvement. It's wise to differentiate between the two in-class policies and syllabi while also explaining what level of AI use is considered appropriate and for which purposes.

WHAT ARE THE LEGALITIES REGARDING COPYRIGHT AND AI USAGE?

The short answer here is that the outcome of the AI copyright battle is TBD. We simply don't know how it's going to end right now. What we DO know is that Amazon, Google, and Microsoft are pouring hundreds of billions of dollars into AI research. Additionally, Amazon controls the self-publishing monopoly and has yet to ban either AI-generated or AI-assisted content. Their terms of service DO require authors to disclose their use of AI, but they will not penalize authors or their books for using the technology.

At the same time, authors are still very much at the mercy of their readers, and bad content is bad content, regardless of who or what generated it, so publishing fully AI-generated content simply isn't a good business move.

This is all the more reason for us as educators to have open conversations with our students about how they can use AI in ways that are contributing something more to society rather than trying to pull the wool over anyone's eyes.

WHAT ARE MOST STUDENTS USING AI FOR?

A recent poll shows that students might not be using AI for all the reasons you'd think (Keierleber).

- Nearly 1/4 use it for general education purposes
- Nearly 1/4 use it as a writing aid
- Nearly 1/3 use it to deal with anxiety or mental health struggles
- Nearly 1/4 use it to work out conflicts with friends
- Nearly 1/5 use it to work out conflicts with family

ARE MOST SCHOOLS DEVELOPING AI POLICIES?

We all know how slow certain school boards and administration teams can move. Although I can't find any evidence that *most* schools are developing AI policies, there are certainly some that have added it to their plagiarism agreements. It seems many of these schools allow AI assistance for brainstorming and writing as long as A) the student does all the major lifting and B) the student cites the AI outputs as appropriate.

WHAT WOULD AN AI POLICY EVEN LOOK LIKE?

Many educators opt for lists to demonstrate which tasks are and are not allowed to be performed by AI. That might look something like this:

Allowed	Not Allowed
Brainstorming and idea generation	Copying and pasting whole sentences
Outlining	Rewriting whole blocks of human text
Asking for high-level suggestions	Passing someone else's work off as your own
Proofreading help	Using hallucinated sources or facts
Reference page formatting	Using AI without citing or disclosing it

Other educators provide sample citations for students to disclose AI use, like this:

OpenAI. (2023). *ChatGPT* (Mar 14 version) [Large language model]. https://chat.openai.com/chat

- ***Parenthetical citation:*** (OpenAI, 2023)
- ***Narrative citation:*** OpenAI (2023)

IS IT BETTER TO ACKNOWLEDGE AI IN THE CLASSROOM OR LET THE STUDENTS DO THEIR OWN RESEARCH?

This question has educators split pretty evenly, and I can't say I blame them. Studies show trust in AI is declining among the older generations while it's doing the opposite among younger ones. However, the general consensus seems to be that whether we talk about it in the classroom, students WILL be interacting with it in some way.

WILL AI ACTUALLY TAKE OVER TEACHING?

If we're all learning from computer overlords in the next 10 years, I will fully eat my words, but at this point in time, I can't imagine a world where teachers go extinct. Learning is a fundamental part of being a human, and mentors make that process a heck of a lot more personable. The robots can't take that away…at least not yet.

Is that a reason to ignore AI and tools like ChatGPT? Not quite. There were likely thousands of teachers who stood up against computers decades ago out of the fear of the unknown, and most of them likely saw tablets and laptops in every single one of their classrooms before they retired.

Many educators also like to point out how poorly remote learning worked out for thousands of students during the onset of the pandemic. Parent involvement is an amazing thing, but teachers, aunts, uncles— you name it—those who are not teachers typically aren't cut out to teach! It's not their fault, but it IS job security.

IS CHATGPT SUITABLE FOR ALL SUBJECTS AND GRADE LEVELS?

You would be hard-pressed to find a single grade level or age group that ChatGPT can't create developmentally appropriate suggestions for.

The model can even develop feeding schedules and sign language lessons for infants who aren't in school and can't even walk yet.

Unsupervised technology use in the younger kiddos isn't suggested with any Internet-based tool, so the same considerations apply here.

WHAT OTHER AI TOOLS ARE OUT THERE BESIDES CHATGPT?

I've included a full list in the Resources section of this book, but the truth is that most (if not all) of these technologies are running off of ChatGPT's systems and knowledge. Some of them will disclose that upfront, but others never do. It's the most comprehensive AI model out there for a reason.

IS IT ETHICAL TO USE AI IN EDUCATION?

I'd like to think the 300+ prompts in this book serve as definitive proof that ethical AI use in education is more than possible, but I'll let you answer that question for yourself.

CHAPTER 16
THE FUTURE OF EDUCATION & NAVIGATING CHANGES WITH AI

"Learning is an ornament in prosperity, a refuge in adversity, and a provision in old age."
—Aristotle, Ancient Greek Philosopher

STUDENTS AREN'T TAKING their jet packs to school or riding flying school buses, but the future of education has already arrived. The US Department of Education recognizes that integrating teaching with AI is not a matter of "if" but "when." They provide the following high-level observations followed by their recommended course of action.

Observations:

1. AI creates new opportunities for communication.
2. AI can increase learning opportunities for different types of learners rather than catering to one ideal.
3. AI makes students and teachers more adaptable.
4. AI allows for deeper and more regular feedback.

5. AI can empower educators to take the reigns over the future of learning.

Recommendations:

1. Maintain a human-first approach to AI training and development.
2. Make sure AI tools are aligned with greater education goals.
3. Keep modern learning principles in mind when developing new tools.
4. Formulate a plan for developing trust in this technology.
5. Keep educators in the loop and let them be involved in the process.
6. Maintain a focus on privacy and data safety measures.
7. Create guidelines that keep students safe.

You'll notice there's no hint of a future without AI because it's simply not probable anymore. The cat's out of the bag, and we have a choice to make. Do we train the models and build them up with our students in mind, or do we let someone else who might not have any teaching experience whatsoever do it for us?

Every time you use ChatGPT to cut down on administrative tasks or create an activity your students have never experienced before, you are paving the pathway for future students and educators, whether you realize it or not.

The next chapter explains how you can take it one step even further if you'd like.

CHAPTER 17
SETTING UP YOUR OWN GPT

"Teaching should be such that what is offered is perceived as a valuable gift and not as hard duty. Never regard study as duty but as the enviable opportunity to learn to know the liberating influence of beauty in the realm of the spirit for your own personal joy and to the profit of the community to which your later work belongs."
—Albert Einstein, Theoretical Physicist

I'M GOING to go ahead and warn you now—this is going to be the most technical chapter out of the entire book, but even then, I've tried my best to make it accessible to even the most technically challenged educators among us.

You can think of this chapter as extra credit. Creating your own GPT is by no means required to begin to enjoy newfound freedoms and inspiration in your classroom, but for those educators who want to go above and beyond, the opportunity is there waiting.

There are three options for creating GPTs: keeping it private to only yourself, sharing it only with those who have the link, or making it fully publicly available. If you choose to do the latter, just know that the full legal name that's attached to your billing account will be made public as well.

What are some reasons for creating your own GPT in the first place? That will depend on who you'd like the model to help and with which tasks. In the earlier chapters, I mentioned creating a virtual tutor of sorts for your students. While that is certainly one noble pathway, you could also create a GPT model entirely dedicated to helping you tackle your email inbox. As a ChatGPT Plus subscriber, you can create unlimited GPT models in as little as 5 minutes, each trained on a unique set of criteria to fulfill a unique need.

This chapter is going to focus on the tutor example, but the steps for creating any GPT are all the same.

Step 1: Log In to Your ChatGPT Account

- Locate your account settings
- Click on the button that says "My GPTs"

Step 2: Select "Create a GPT"

Step 3: Begin to Train Your GPT Model

- You will summarize what you want your new GPT to do in one sentence, such as "Help tutor 5th-grade students"
- You will be able to customize the GPT description, name, and even photo
- ChatGPT will ask you questions about how you want your model to interact with your audience, prompting you to give it more information about the style and tone you have in mind
- A split view screen will appear with your GPT editor on the left and the model that you are building on the right

- You can play around with the model on the right and keep giving the editor more parameters on the left until you are happy with the outputs it is creating
- Consider adding safety filters if you plan on having students interact with the model

Step 4: Set Up Access

- Choose whether you'd like your GPT model to be private, open to only those with a link, or fully open to the public

Step 5: Gather Feedback and Adjust

- Users of your GPT model will be able to leave feedback about their experience
- Use this feedback to adjust your model as needed, i.e., tweaking accessibility features

GPT IDEAS

- **Homework Helper:** Students can ask specific questions about their homework and receive guided explanations or hints.
- **Study Buddy:** For exam preparation, students can quiz themselves on various topics, and ChatGPT will provide answers and explanations.
- **Writing Assistant:** Students working on essays or reports can use ChatGPT to brainstorm, structure their work, or refine their arguments.
- **Journaling:** Students can generate unlimited journal prompts to help them work on their self-confidence and emotional regulation.

IMPORTANT CONSIDERATIONS

- **Data Privacy**: Adhere to COPPA (Children's Online Privacy Protection Act) and FERPA (Family Educational Rights and Privacy Act) guidelines to protect student information.
- **Content Accuracy**: Double-check the outputs you receive from any GPT model before assuming they are accurate.
- **Equitable Access**: Make sure every student in your class can access ChatGPT, including those with limited internet access or disabilities.

RESOURCES FOR TEACHERS

I've compiled a list of free teacher resources out there for those who wish to explore other avenues for integrating AI into the classroom. I will be updating this list with every future iteration of the book.

AI Tools and Platforms for Education

- AudioPen: A voice-to-text tool
- Canva: A graphic design and presentation design tool
- Google AI Offers various AI tools and educational resources, including machine learning crash courses and AI experiments.
- Kahoot!:An AI-based platform that makes learning fun with quizzes and interactive games.
- Otter.ai: A meeting and transcription wizard
- Quillionz: Uses AI to generate quiz questions and learning materials from the text.
- Socratic by Google: An AI-powered app that helps students with homework by providing explanations, videos, and definitions.

- Zzish: A platform that uses AI to create personalized learning experiences and provide real-time insights into student understanding.

Educational Content and Curriculum

- Magic School: A lesson plan creator
- MIT OpenCourseWare: Offers free course materials on AI and related subjects from one of the leading institutions in AI research.
- Coursera: Provides courses on AI and machine learning for educators, including "AI For Everyone" by Andrew Ng, which is suitable for non-technical learners.
- Curipod: An interactive lesson plan creator
- Eduaide: An AI-assisted lesson plan, assessment, and resource creator
- EdX: Offers courses and professional certificates in AI, machine learning, and data science from universities like Harvard and MIT.
- Khan Academy: Although not AI-focused, it offers foundational courses in computer science and mathematics that are essential for understanding AI.
- Grammarly: Offers writing, editing, and proofreading assistance
- Scribbr: Offers citation assistance.
- Slidesgo: A presentation creator

Professional Development and Teacher Training

- ISTE U: The International Society for Technology in Education offers courses on AI in education, helping teachers understand how to apply AI in teaching.
- Teachable Machine by Google: A tool that allows teachers and students to create machine learning models without coding, making AI concepts accessible and understandable.

- Microsoft Educator Center: Offers training and resources for educators, including content on integrating AI tools into teaching and learning.
- AI4ALL Open Learning: Provides free, accessible AI education materials aimed at high school teachers and students, focusing on ethical implications and diverse applications of AI.
- AI4K12.org: Free training for teachers on integrating AI into K-12 classrooms.

Online Communities and Forums

- EdSurge: An educational technology community that discusses the latest in education innovation, including AI applications in education.
- -LinkedIn and Facebook Groups: There are various groups dedicated to educational technology and AI in education where teachers can share experiences, ask questions, and find support.

Journals and Publications

- American Association of Colleges and Universities: Offers insights regarding all the latest teaching initiatives, including AI.
- International Journal of Artificial Intelligence in Education: Offers research papers and articles on the latest AI applications in education.
- The Journal of AI Research: Provides access to the latest research in AI, some of which can be applied in educational contexts.

AFTERWORD

As we come to the end of **The AI Chatbot for Teachers**, I hope you have everything you need to use AI in your classroom in a creative and useful way. This is the book you need to learn how to make your lessons more fun and less dull for your schoolmates.

One way to look at artificial intelligence (AI) is as a helpful classroom assistant that gives you more time to teach and get to know your students, which are two things you really enjoy. Also, remember that you are just starting out on this journey; there is a lot you can learn and improve.

Share your thoughts and questions, and let's keep pushing the edges of what technology can do in the classroom.

It's great to have you along for the ride. Yes, to the exciting journey ahead that will teach you lots of new things!

Sheila Sonne

ABOUT THE AUTHOR

Sheila Sonne has been a teacher for more than 20 years and is also a published author. Her book, The AI Chatbot for Teachers: 300+ Instant Chat Prompts to Engage Students, Simplify Lessons, and Cut Down on Admin Work, focuses on her knowledge in this area. Teachers may utilize this handbook to better engage students, leverage AI to increase learning, and minimize administrative work.

Sheila's research focuses on the practical applications of artificial intelligence in education, with the goal of increasing teaching efficiency and improving student results. Her collaborations with educational technology businesses also help to build classroom-specific software solutions. Sheila is at the forefront of researching and writing on how adaptive learning technology may customize and enhance education.

WORKS CITED

Centers for Disease Control and Prevention. "Data and Statistics about ADHD." *Centers for Disease Control and Prevention*, 9 Aug. 2022, www.cdc.gov/ncbddd/adhd/data.html.

"COE - Teachers' Reports on Managing Classroom Behaviors."*Condition of Education*, U.S. Department of Education, Institute of Education Sciences, 2023, nces.ed.gov/programs/coe/indicator/a11.

Coffey, Lauren. "Students Outrunning Faculty in AI Use." *Inside Higher Ed*, 31 Oct. 2023, www.insidehighered.com/news/tech-innovation/artificial-intelligence/2023/10/31/most-students-outrunning-faculty-ai-use.

Fensterwald, John. "Latest National Test Results Underscore Declining Knowledge of U.S. History and Civics." *EdSource*, 3 May 2023, edsource.org/2023/latest-test-results-underscore-declining-knowledge-of-u-s-history-and-civics/689766#:~:text=Only%2013%25%20of%20students%20scored.

Golden Steps ABA. "49 Reading Statistics & Facts You Should Know." *www.goldenstepsaba.com*, 31 July 2023, www.goldenstepsaba.com/resources/reading-statistics#:~:text=A%20survey%20conducted%20by%20Scholastic.

Hamilton, Ilana, and Brenna Swanston. "Artificial Intelligence in the Classroom: What Do Educators Think? – Forbes Advisor." *Forbes*, 5 Dec. 2023, www.forbes.com/advisor/education/it-and-tech/artificial-intelligence-in-school/#:~:text=60%25%20of%20Educators%20Use%20AI.

Hutton, Lisa, et al. *The State of K-12 History Teaching: Challenges to Innovation | Perspectives on History | AHA*. American Historical Association, 1 May 2012, www.historians.org/research-and-publications/perspectives-on-history/may-2012/the-state-of-k-12-history-teaching.

IBM. "What Is Artificial Intelligence (AI)?" *IBM*, 2023, www.ibm.com/topics/artificial-intelligence.

Indeed Editorial Team. "Why Is Instant Feedback Important? (plus 6 Benefits)." *Indeed*, 30 Sept. 2022, www.indeed.com/career-advice/career-development/why-is-instant-feedback-important.

Keierleber, Mark. "ChatGPT Is Landing Kids in the Principal's Office, Survey Finds." *The74*, 20 Sept. 2023, www.the74million.org/article/chatgpt-is-landing-kids-in-the-principals-office-survey-finds/?utm_source=pocket_saves.

Levitan, Shayna. "Planning Time May Help Mitigate Teacher Burnout—but How Much Planning Time Do Teachers Get?" *National Council on Teacher Quality (NCTQ)*, 12 Jan. 2023, www.nctq.org/blog/Planning-time-may-help-mitigate-teacher-burnoutbut-how-much-planning-time-do-teachers-get.

May, Michael E. "Effects of Differential Consequences on Choice Making in Students at Risk for Academic Failure." *Behavior Analysis in Practice*, vol. 12, no. 1, 31 May

2018, pp. 154–161, www.ncbi.nlm.nih.gov/pmc/articles/PMC6411561/, https://doi.org/10.1007/s40617-018-0267-3.

McShane, Michael. *How Do Teachers Spend Their Time?* edChoice, July 2022.

National Academies. "Science Education Should Be National Priority; New Report Calls on Federal Government to Encourage Focusing Resources on High-Quality Science for All Students." *National Academies*, 13 July 2021, www.nationalacademies.org/news/2021/07/science-education-should-be-national-priority-new-report-calls-on-federal-government-to-encourage-focusing-resources-on-high-quality-science-for-all-students#:~:text=Only%2022%20percent%20of%20American.

National Center for Education Statistics. "Characteristics of 2020–21 Public and Private K–12 Schools in the United States: Results from the National Teacher and Principal Survey." *Nces.ed.gov*, National Center for Education Statistics, 13 Dec. 2022, nces.ed.gov/pubsearch/pubsinfo.asp?pubid=2022111.

"More than 80 Percent of U.S. Public Schools Report Pandemic Has Negatively Impacted Student Behavior and Socio-Emotional Development." *National Center for Education Statistics*, 6 July 2022, nces.ed.gov/whatsnew/press_releases/07_06_2022.asp.

Office of Educational Technology. "Artificial Intelligence." *Office of Educational Technology*, 2023, tech.ed.gov/ai/.

Pearce, Allie. "Fact Sheet: 3 Trends in K-12 Assessments across the Country." *Center for American Progress*, 11 Jan. 2024, www.americanprogress.org/article/fact-sheet-3-trends-in-k-12-assessments-across-the-country/.

Richards, Erin. "Math Scores Stink in America. Other Countries Teach It Differently - and See Higher Achievement." *USA TODAY*, 29 Feb. 2020, www.usatoday.com/story/news/education/2020/02/28/math-scores-high-school-lessons-freakonomics-pisa-algebra-geometry/4835742002.

Sparks, Sarah D. "The State of Math Education, in Charts." *Education Week*, 31 July 2023, www.edweek.org/teaching-learning/the-state-of-math-education-in-charts/2023/07.

Stone, Peter, et al. "Artificial Intelligence and Life in 2030." One Hundred Year Study on Artificial Intelligence: Report of the 2015-2016 Study Panel, *Stanford University*, Sept. 2016, http://ai100.stanford.edu/2016-report.

U.S. Department of Education, Office of Educational Technology, Artificial Intelligence and Future of Teaching and Learning: Insights and Recommendations, Washington, DC, 2023.

Wexler, Natalie. "Why Kids Know Even Less about History Now—and Why It Matters." *Forbes*, 24 Apr. 2020, www.forbes.com/sites/nataliewexler/2020/04/24/why-kids-know-even-less-about-history-now-and-why-it-matters/?sh=b7b25676a7af.

Zapato, Lyle. Save The Pacific Northwest Tree Octopus. United States, 2011. Web Archive. Retrieved from the Library of Congress, <www.loc.gov/item/lcwaN0010826/>.